METHODS IN MOLECULAR BIOLOGY

Series Editor
John M. Walker
School of Life and Medical Sciences
University of Hertfordshire
Hatfield, Hertfordshire, UK

For further volumes:
http://www.springer.com/series/7651

For over 35 years, biological scientists have come to rely on the research protocols and methodologies in the critically acclaimed *Methods in Molecular Biology* series. The series was the first to introduce the step-by-step protocols approach that has become the standard in all biomedical protocol publishing. Each protocol is provided in readily-reproducible step-by-step fashion, opening with an introductory overview, a list of the materials and reagents needed to complete the experiment, and followed by a detailed procedure that is supported with a helpful notes section offering tips and tricks of the trade as well as troubleshooting advice. These hallmark features were introduced by series editor Dr. John Walker and constitute the key ingredient in each and every volume of the *Methods in Molecular Biology* series. Tested and trusted, comprehensive and reliable, all protocols from the series are indexed in PubMed.

Rice

Methods and Protocols

Edited by

Taiji Kawakatsu

Institute of Agrobiological Sciences, National Agriculture and Food Research Organization, Tsukuba, Ibaraki, Japan

Humana Press

Editor
Taiji Kawakatsu
Institute of Agrobiological Sciences
National Agriculture and Food Research Organization
Tsukuba, Ibaraki, Japan

ISSN 1064-3745 ISSN 1940-6029 (electronic)
Methods in Molecular Biology
ISBN 978-1-0716-4203-0 ISBN 978-1-0716-4204-7 (eBook)
https://doi.org/10.1007/978-1-0716-4204-7

This Humana imprint is published by the registered company Springer Science+Business Media, LLC, part of Springer Nature.
The registered company address is: 1 New York Plaza, New York, NY 10004, U.S.A.

Preface

Rice is Life

Rice is one of the three major crops and accounts for 17% of the global food supply on a calorie basis in 2021. With the global population projected to surpass 9 billion by 2050, it is imperative that annual cereal production increases by 1.5-fold to meet this demand. Furthermore, the impact of climate change on grain yield necessitates the development of crops resilient to such changes.

Rice, with its compact and simple diploid genome, has been a widely studied model cereal crop. The advance of high-throughput sequencing and genome editing technologies over the past two decades has revolutionized the field. Resequencing thousands of rice genomes has provided invaluable resources for understanding genomic diversity, tracing the path of domestication, and studying evolution. Transcriptome analyses offer snapshots of cellular states, shedding light on various biological processes. Genome editing has empowered researchers to generate mutants for specific genes of interest. These advancements are integral to the field of molecular biology, particularly in accelerating functional genomics. This, in turn, paves the way for molecular breeding strategies that could significantly contribute to achieving the Sustainable Development Goals (SDGs).

The aim of this volume is to serve as a comprehensive guide to the field of rice molecular biology, encompassing both time-honored techniques and state-of-the-art methodologies. This book is divided into six parts: sample preparation, temporal and spatial analysis of RNA and protein expression, transformation and precise genome editing, scalable transcriptome analysis, epigenome analysis, and gene regulatory network analysis. They hold the promise of unlocking novel findings in rice science, paving the way for innovative solutions to global food security challenges.

All the authors and I hope that this volume will serve as a valuable resource for the rice research community and beyond. Our collective knowledge is our most potent weapon in the fight for global food security. We envision a future where these tools and techniques are widely adopted, driving advancements in crop science and bringing us one step closer to a world where food security is a reality for all.

Tsukuba, Ibaraki, Japan *Taiji Kawakatsu*

Contents

Preface . . . . *v*
Contributors . . . . *ix*

1 Hydroponic Culture of Rice Seedlings for Stress Response Assay 1
Fumiyuki Soma and Yusaku Uga

2 Total RNA Extraction from Rice Vegetative Tissues Using Magnetic Beads 7
Kanami Yoshino and Taiji Kawakatsu

3 Total RNA Extraction from Starchy Rice Endosperm 15
Yuhya Wakasa

4 RNA Extraction from Rice Immature Embryo Using Laser Capture Microdissection 21
Takumi Tezuka, Misuzu Nosaka-Takahashi, Takanori Yoshikawa, and Yutaka Sato

5 Using Immunoblotting for Detection of a Circadian Clock Component, Os–GI, in Rice Leaves 29
Hironori Itoh

6 Preparation and Sectioning of Paraffin-Embedded Tissue for Histology and Histochemistry 41
Masayuki Miya, Ken-ichiro Hibara, and Jun-ichi Itoh

7 Spatial Analysis of Gene Expression by In Situ Hybridization 49
Masayuki Miya, Ken-ichiro Hibara, and Jun-ichi Itoh

8 Purification of Anti-OSH1 IgG and Immunohistochemical Staining of Rice Shoot Apical Meristem 61
Takumi Tezuka, Misuzu Nosaka-Takahashi, Takanori Yoshikawa, and Yutaka Sato

9 *Agrobacterium*-Mediated Transformation and Targeted Mutagenesis Using SpCas12f in Rice 75
Satoru Sukegawa, Seiichi Toki, and Hiroaki Saika

10 Gene Targeting via CRISPR/Cas9-Mediated DNA Double-Strand Break Induction in Rice 91
Ayako Nishizawa-Yokoi and Seiichi Toki

11 Precise Base Substitution Using CRISPR/Cas-Mediated Base Editor in Rice 101
Masaki Endo, Katsuya Negishi, and Seiichi Toki

12 DeLTa-Seq: High-Throughput Targeted RNA-Seq of Rice Leaves Without RNA Purification 113
Makoto Kashima, Yasuyuki Nomura, and Atsushi J. Nagano

13 Lasy-Seq: A High-Throughput 3′ RNA-Seq Method for Large-Scale Transcriptome Profiling in Rice 123
Natsumi Mori-Moriyama and Atsushi J. Nagano

14 Profiling of 24-nt phasiRNAs in Rice Anther by Massively Parallel Sequencing and Stem-Loop Reverse Transcription PCR 135
Huong Ta and Ken-Ichi Nonomura

15 Expression and Purification of pAG-MNase for CUT&RUN................ 147
Taiji Kawakatsu

16 Chromatin Immunoprecipitation from Shoot Apices and Young Panicles of Rice.. 157
Katsutoshi Tsuda

17 In Vitro Genome-Wide Identification of Transcription Factor Binding Sites 169
Taiji Kawakatsu

18 High-Efficient Transient Gene Transfection for Transcriptional Activity Assay Using Rice Protoplasts on Large-Scale Platform 183
Shingo Sakamoto

Index *193*

Contributors

MASAKI ENDO • *Institute of Agrobiological Sciences, National Agriculture and Food Research Organization, Tsukuba, Ibaraki, Japan*
KEN-ICHIRO HIBARA • *Graduate School of Agricultural Regional Vitalization, Kibi International University, Minamiawaji, Japan*
HIRONORI ITOH • *Institute of Crop Science, National Agriculture and Food Research Organization, Tsukuba, Ibaraki, Japan*
JUN-ICHI ITOH • *Graduate School of Agricultural and Life Sciences, University of Tokyo, Tokyo, Japan*
MAKOTO KASHIMA • *Faculty of Science, Department of Molecular Biology, Toho University, Chiba, Japan*
TAIJI KAWAKATSU • *Institute of Agrobiological Sciences, National Agriculture and Food Research Organization, Tsukuba, Ibaraki, Japan*
MASAYUKI MIYA • *National Institute of Genetics, Mishima, Shizuoka, Japan*
NATSUMI MORI-MORIYAMA • *Faculty of Agriculture, Ryukoku University, Shiga, Japan*
ATSUSHI J. NAGANO • *Faculty of Agriculture, Ryukoku University, Shiga, Japan; Institute for Advanced Biosciences, Keio University, Yamagata, Japan*
KATSUYA NEGISHI • *Institute of Fruit Tree and Tea Science, National Agriculture and Food Research Organization, Tsukuba, Ibaraki, Japan*
AYAKO NISHIZAWA-YOKOI • *Institute of Agrobiological Sciences, National Agriculture and Food Research Organization, Tsukuba, Ibaraki, Japan*
YASUYUKI NOMURA • *Research Institute for Food and Agriculture, Ryukoku University, Shiga, Japan*
KEN-ICHI NONOMURA • *Plant Cytogenetics Laboratory, National Institute of Genetics, Mishima, Shizuoka, Japan; Genetics Course, The Graduate University for Advanced Studies (SOKENDAI), Mishima, Shizuoka, Japan*
MISUZU NOSAKA-TAKAHASHI • *National Institute of Genetics, Mishima, Shizuoka, Japan*
HIROAKI SAIKA • *Institute of Agrobiological Sciences, National Agriculture and Food Research Organization, Tsukuba, Ibaraki, Japan*
SHINGO SAKAMOTO • *Bioproduction Research Institute, National Institute of Advanced Industrial Science and Technology (AIST), Tsukuba, Ibaraki, Japan; Global Zero-Emission Research Center, National Institute of Advanced Industrial Science and Technology (AIST), Tsukuba, Ibaraki, Japan*
YUTAKA SATO • *National Institute of Genetics, Mishima, Shizuoka, Japan*
FUMIYUKI SOMA • *Institute of Crop Science, National Agriculture and Food Research Organization, Tsukuba, Ibaraki, Japan*
SATORU SUKEGAWA • *Institute of Agrobiological Sciences, National Agriculture and Food Research Organization, Tsukuba, Ibaraki, Japan*
HUONG TA • *Plant Cytogenetics Laboratory, National Institute of Genetics, Mishima, Shizuoka, Japan*
TAKUMI TEZUKA • *National Institute of Genetics, Mishima, Shizuoka, Japan*

SEIICHI TOKI • *Institute of Agrobiological Sciences, National Agriculture and Food Research Organization, Tsukuba, Ibaraki, Japan; Graduate School of Nanobioscience, Yokohama City University, Kanagawa, Japan; Kihara Institute for Biological Research, Yokohama City University, Kanagawa, Japan; Department of Life Science, Faculty of Agriculture, Ryukoku University, Kyoto, Japan*
KATSUTOSHI TSUDA • *Plant Cytogenetics Laboratory, Department of Gene Function and Phenomics, National Institute of Genetics, Mishima, Shizuoka, Japan; Department of Genetics, School of Life Science, Graduate University for Advanced Studies, Mishima, Shizuoka, Japan*
YUSAKU UGA • *Institute of Crop Science, National Agriculture and Food Research Organization, Tsukuba, Ibaraki, Japan*
YUHYA WAKASA • *Institute of Agrobiological Sciences, National Agriculture and Food Research Organization, Tsukuba, Ibaraki, Japan*
TAKANORI YOSHIKAWA • *National Institute of Genetics, Mishima, Shizuoka, Japan*
KANAMI YOSHINO • *Department of Agrobiology and Bioresources, Faculty of Agriculture, Utsunomiya University, Utsunomiya, Tochigi, Japan*

Chapter 1

Hydroponic Culture of Rice Seedlings for Stress Response Assay

Fumiyuki Soma and Yusaku Uga

Abstract

The major environmental factors limiting rice growth and production are osmotic stresses such as drought and high salinity. High osmotic stresses directly disrupt cellular activities, leading to plant growth retardation or death. Plants have various response mechanisms to survive under such stresses. Understanding rice's stress response mechanisms is necessary to enhance the osmotic stress tolerance of rice. However, assessing specific physiological responses to osmotic stresses is difficult because multiple environmental factors affect rice growth. Here, we describe a simple method for analyzing the osmotic stress responses of rice plants using a hydroponic culture system. This method allows comprehensive gene expression and phenotypic analyses under osmotic stress conditions in rice. Various osmotic stress conditions and samples can be tested simultaneously because this method is small-scale. In addition, the procedure is easy, and highly reproductive results can be obtained.

Key words Osmotic stress, Drought stress, Gene expression, Phenotyping

1 Introduction

Osmotic stresses, such as drought and high salinity, negatively affect rice growth and productivity [1–3]. Plants have various osmotic stress response mechanisms to survive. The expressions of many genes encoding transcription factors, transporters, chaperons, or hydrophilic proteins are induced in response to osmotic stress to acquire stress tolerance [4]. Identifying and characterizing genes associated with osmotic stress response mechanisms is needed to develop stress-tolerant rice plants [5]. However, the gene expression levels are sensitive to various environmental factors, making it challenging to evaluate specific environmental responses accurately. Some methods for assessing the osmotic stress responses of rice using hydroponic culture have been developed [6, 7]. Here, we described a simple method for evaluating osmotic

Taiji Kawakatsu (ed.), *Rice: Methods and Protocols*, Methods in Molecular Biology, vol. 2869,
https://doi.org/10.1007/978-1-0716-4204-7_1,

stress responses at the seedling stage using a hydroponic culture system. This method allows for assessing gene expression under osmotic stress and protein level or growth changes in chemical or plant hormone treatment. Compared to previous hydroponic methods, this method is small-scale and straightforward, allowing for reproducible analysis of large samples and stress conditions.

2 Materials

Prepare all solutions using ultrapure water (18.2 MΩ cm at 25 °C, e.g., Milli-Q water) and analytical grade reagents (*see* **Note 1**).

1. Rice seeds (e.g., *Oryza sativa*, japonica cultivar Nipponbare seeds).
2. Pesticides (e.g., pefurazoate) or biocide/fungicide (e.g., Plant Preservative Mixture).
3. Growth chamber: Set 14 h light/10 h dark photoperiod, 25–30 °C and 40–60% humidity.
4. Modified Kimura B stock solution I: Add 24.1 g of $(NH_4)_2SO_4$, 7.9 g of K_2SO_4, 67.4 g of $MgSO_4{\bullet}7H_2O$, 9.3 g of KNO_3, and 12.4 g of KH_2PO_4 to a 1 L autoclavable bottle with screw cap containing approximately 900 mL ultrapure water. Make up to 1 L with ultrapure water. Autoclave and store at room temperature (*see* **Note 2**).
5. Modified Kimura B stock solution II: Add 2 g of $FeC_6H_5O_7{\bullet}nH_2O$ to 1 L autoclavable bottle with screw cap containing 500 mL of hot 0.2 N HCl (90 °C). After dissolution, add 43.1 g of $Ca(NO_3)_2{\bullet}4H_2O$. Make up to 1 L with ultrapure water. Autoclave and store at room temperature in the dark (*see* **Note 2**).
6. Hydroponic solution: 1 mL/L modified Kimura B stock solution I, 1 mL/L modified Kimura B stock solution II. Adjust to pH 5.5 with 1 M potassium hydroxide. Prepare fresh before use.
7. Cultivation net: Cut 5 mm (thickness) styrene board into 60 × 60 (mm) squares and drill a 40 × 40 (mm) square hole in the center (*see* Fig. 1a). Cut 1 mm nylon mesh into 50 × 50 (mm) squares (*see* Fig. 1b). Attach a nylon mesh at the bottom of the styrene board using a stapler (*see* Fig. 1c).
8. 60 × 60 × 100 (width × depth × height, mm) plastic pot (*see* Fig. 1d) or other larger pot.
9. PEG solution: 0, 10, 20, and 30% (w/v) polyethylene glycol 6000 in hydroponic solution.

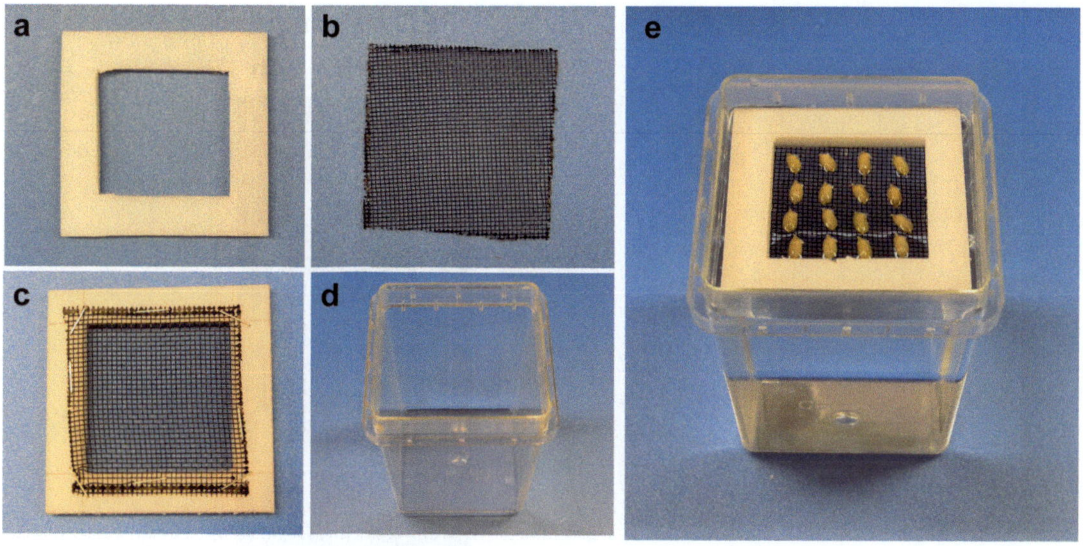

Fig. 1 Hydroponic culture of rice. (**a**) Cut styrene board. (**b**) Cut nylon mesh. (**c**) Cultivation net. (**d**) Plastic pot. (**e**) Hydroponic cultivation system after transferring the germinated rice seeds

3 Methods

3.1 Seed Germination and Plant Growth

1. Sterilize the surface of rice seeds with pesticides or biocide/fungicide in the dark at 15 °C for 24 h (*see* **Note 3**).
2. Soak the seeds in water and incubate in the dark at 30 °C for 2 days.
3. Add 300 mL of hydroponic solution into a plastic pot.
4. Float a cultivation net on the hydroponic solution.
5. Place the germinated rice seeds from **step 2** onto a cultivation net floated in hydroponic solution in a plastic pot (*see* Fig. 1e and **Note 4**).
6. Incubate in a growth chamber (14 h light at 30 °C and 10 h dark at 25 °C, 50% relative humidity).
7. Replace the hydroponic solution every 2–7 days (*see* **Note 5**).

3.2 Osmotic Stress Treatment

1. After 8–14 days, transfer the cultivation net with rice plants from the hydroponic solution to the PEG solution (*see* **Notes 6** and 7).
2. Collect root or shoot tissues to analyze early response to osmotic stress after 1–24 h of PEG treatment. These samples can be used as is for transcriptome or proteome analysis.
3. After 1–3 days of PEG treatment, observe the plant phenotype under osmotic stress conditions compared with its control conditions (*see* Fig. 2 and **Note 8**).

Fig. 2 Osmotic stress treatments using a hydroponic culture system. Rice seedlings at 12 days after transplanting subjected to PEG treatment for 0, 1, and 24 h. Leaf rolling and wilting were observed under 30% PEG condition

4 Notes

1. Ultrapure water is recommended for the preparation of all solutions to improve reproducibility.
2. We usually use Kimura B-based hydroponic solution [8], but other hydroponic solutions, such as Murashige–Skoog liquid medium, can be used.
3. We use 375 ppm of pefurazoate or 1% (v/v) Plant Preservative Mixture [9, 10]. Alternatively, sterilize the surface of rice seeds with 1–5% sodium hypochlorite solution for 10–30 min, followed by washing with tap water three times.
4. Up to 16 plants can be grown per pot (*see* Fig. 1).
5. The hydroponic solution decreases due to transpiration. It depends on the growth conditions (e.g., humidity) and the rice variety.
6. The optimal growth condition and stress treatment vary across rice varieties. For example, Kinandang Patong was more sensitive to osmotic stress than IR64 [1]. Therefore, it is crucial to conduct optimization for each specific rice variety.

7. In addition to other stress conditions such as high salinity or heat stress, plant hormone or chemical treatment can be applicable [1, 2]. Because solvent (e.g., DMSO) may affect plant growth, it is necessary to titrate the solvent concentration and compare the growth between conditions with no additives and those with a mock treatment (solvent only).
8. Root and shoot length, fresh weight, leaf rolling, and wilting can be evaluated (*see* Fig. 2). Leaf rolling was usually observed after 1 h of 30% PEG treatment.

Acknowledgments

This work was financially supported by JST CREST (grant number JPMJCR17O1) and a JSPS Grant-in-Aid for Young Scientists (22 K14864).

References

1. Uga Y, Sugimoto K, Ogawa S et al (2013) Control of root system architecture by DEEPER ROOTING 1 increases rice yield under drought conditions. Nat Genet 45:1097–1102
2. Kitomi Y, Hanzawa E, Kuya N et al (2020) Root angle modifications by the DRO1 homolog improve rice yields in saline paddy fields. Proc Natl Acad Sci USA 117:21242–21250
3. Soma F, Kitomi Y, Kawakatsu T et al (2024) Life-cycle multiomics of rice shoots reveals growth stage-specific effects of drought stress and time–lag drought responses. Plant Cell Physiol 65:156–168
4. Soma F, Takahashi F, Yamaguchi-Shinozaki K et al (2021) Cellular phosphorylation signaling and gene expression in drought stress responses: ABA-dependent and ABA-independent regulatory systems. Plan Theory 10:756
5. Bailey-Serres J, Parker JE, Ainsworth EA et al (2019) Genetic strategies for improving crop yields. Nature 575:109–118
6. Cabuslay GS, Ito O, Alejar A (2002) Physiological evaluation of responses of rice (Oryza sativa L.) to water deficit. Plant Sci 163:815–827
7. Kato Y, Hirotsu S, Nemoto K et al (2008) Identification of QTLs controlling rice drought tolerance at seedling stage in hydroponic culture. Euphytica 160:423–430
8. Yoshida S, Forno D, Cock J et al (1976) Laboratory manual for physiological studies of rice. IRRI, Philippines
9. Miyoshi Y, Soma F, Yin YG et al (2023) Rice immediately adapts the dynamics of photosynthates translocation to roots in response to changes in soil water environment. Front Plant Sci 13:1024144
10. Numajiri Y, Yoshino K, Teramoto S et al (2021) iPOTs: internet of things-based pot system controlling optional treatment of soil water condition for plant phenotyping under drought stress. Plant J 107:1569–1580

Chapter 2

Total RNA Extraction from Rice Vegetative Tissues Using Magnetic Beads

Kanami Yoshino and Taiji Kawakatsu

Abstract

Ribonucleic acid (RNA) extraction is the first critical step in gene expression analysis. In this chapter, we describe a high-throughput RNA extraction method using guanidine thiocyanate and isopropyl alcohol (HighGI). The use of carboxyl-coated paramagnetic beads, instead of silica membrane columns, enables semi-automation using a liquid handling system and high-throughput RNA extraction for large-scale transcriptome studies. Homemade mixes of paramagnetic beads and buffers make HighGI inexpensive. In addition, HighGI-extracted RNA retains low molecular weight RNA molecules less than 200 bp, which is typically lost in commercial column-based kits.

Key words Gene expression analysis, Transcriptomes, Guanidine thiocyanate, Magnetic beads, Small RNA extraction

1 Introduction

RNAs are the transcripts of genes, and their composition and abundance are indicators of the activity of biological processes such as development, metabolism, and response to environmental stimuli. RNA extraction is an essential procedure for downstream gene expression analyses such as reverse transcription quantitative PCR, microarray, and RNA-seq to understand the current state of tissues or cells. Advances in high-throughput sequencing technology have made it possible to acquire over 1000 samples in 2 days, but sample preparation steps such as RNA extraction and library preparation have become bottlenecks in terms of cost and labor, especially when analyzing a huge number of samples. Classic methods of RNA extraction include phenol/chloroform extraction and acid–guanidine thiocyanate (GTC)–phenol–chloroform extraction [1–3]. However, phenol is corrosive and toxic and requires special attention because of its health risks. RNA extraction procedures based on cetyltrimethylammonium bromide (CTAB) and

Taiji Kawakatsu (ed.), *Rice: Methods and Protocols*, Methods in Molecular Biology, vol. 2869,
https://doi.org/10.1007/978-1-0716-4204-7_2,

polyvinylpyrrolidone (PVP) have also been employed, especially for tissues with high secondary metabolites such as polysaccharides and polyphenols [4–7]. RNA extraction with CTAB and PVP is useful for viscous materials but is limited by the number of tubes that can be centrifuged in parallel because it requires ethanol or isopropyl alcohol (IPA) precipitation and lengthy centrifugation processes. GTC is a deproteinizing agent [8] and effectively inhibits RNase activity. The combination of GTC and reducing agents such as dithiothreitol or 2-mercaptoethanol is one of the most commonly used RNA extraction methods [2, 9, 10]. Many commercial kits use GTC in lysis and wash buffers to denature proteins. GTC is a chaotropic salt, and the high concentration of GTC enhances binding of nucleic acids to silica membrane columns reversibly. Therefore, GTC- and column-based commercial kits can omit time-consuming ethanol precipitation steps, allowing rapid RNA extraction. However, commercial kits are expensive, and column-based kits are unsuitable for processing a large number of samples. Solid-phase reversible immobilization (SPRI) beads, which are silica- or carboxyl-coated paramagnetic beads, reversibly bind to nucleic acids in the presence of the macromolecular crowding agent polyethylene glycol and a salt. SPRI beads are used for the low-cost and high-throughput preparation of DNA sequencing libraries [11]. While ethanol is often added to the lysis buffer homogenates in column-based commercial kits to promote the binding of hydrophilic nucleic acids to the silica membrane, IPA can precipitate nucleic acids more effectively than ethanol at lower concentrations. In this chapter, we introduce a high-throughput GTC- and IPA-based RNA extraction method (HighGI RNA extraction method) that utilizes carboxyl-coated paramagnetic beads and can be semi-automated using a liquid handling system. The yield, purity, and integrity of RNA extracted by the HighGI are comparable to those by conventional phenol/chloroform-based methods and column-based commercial kits [12]. Addition of IPA, instead of ethanol, to the lysis buffer homogenates retains lower molecular weight RNAs, including tRNAs and small RNAs, which are lost in column-based commercial kits. Homemade mixes of paramagnetic beads and buffers can be easily prepared. The HighGI method is inexpensive, versatile, and suitable for large-scale transcriptome studies.

2 Materials

Prepare all solutions using ultrapure nuclease-free distilled water and analytical grade reagents.

2.1 Plant Materials

We successfully extract RNA from various rice tissues, including young shoots and leaves with fresh weights of up to 50 mg and five individual root tips (*see* **Notes 1** and **2**).

2.2 Equipment

1. Bead mill homogenizer or mortar and pestle (*see* **Note 3**).
2. Magnetic stand for 1.5 mL tubes.
3. NanoDrop spectrophotometer.
4. Agilent 2100 Bioanalyzer (optional; *see* **Note 4**).
5. Agilent RNA 6000 Nano/Pico Kit (optional; *see* **Note 4**).

2.3 Reagents and Solutions

1. Guanidine thiocyanate (hereafter GTC).
2. Isopropyl alcohol (hereafter IPA).
3. TE buffer: 10 mM Tris–HCl (pH 8.0), 10 mM EDTA. Autoclave and store at room temperature.
4. 85% (v/v) ethanol.
5. Recombinant DNase I (optional; *see* **Note 5**).
6. Lysis buffer: 1 mM Tris–HCl (pH 7.5), 4 M GTC, 0.05% (v/v) Tween 20. Store the mixture at 4 °C. Lysis buffer is stable for 1 month.
7. SPRI beads: AMPure XP or prepare as follows. Transfer 1 mL of suspended Sera-Mag Carboxylate-modified SpeedBeads to a 1.5 mL microtube. Place the tube on a magnetic stand and wait until the beads are settled on the magnet. Discard supernatant and remove the tube from the magnetic stand. Resuspend beads in 1 mL of TE, place it back to the magnetic stand, and remove supernatant. Repeat TE wash twice to remove sodium azide included in Sera-Mag Carboxylate-modified SpeedBeads. Resuspend beads in 1 mL of TE. In a new 50 mL conical tube, add 10 g of PEG 8000, 25 mL of 5 M NaCl, 500 μL of 1 M Tris–HCl (pH 8.0), and 100 μL of 0.5 M EDTA (pH 8.0), and fill the conical tube with ultrapure water up to 45 mL line. Mix the tube on a rotary shaker until PEG is dissolved. Add 250 μL of 10% (v/v) Tween 20 and mix gently. Add 1 mL of TE suspended Sera-Mag SpeedBeads. Fill the conical tube with nuclease-free ultrapure water up to 50 mL and mix gently. Store in the dark or wrap in aluminum foil at 4 °C. Homemade SPRI beads are stable for at least 3 months.
8. Wash buffer: 1 mM Tris–HCl, pH 7.5, 1 M GTC, 40% (v/v) IPA, 1 M NaCl, 0.05% (v/v) Tween 20.

2.4 Semi-Automation (Optional; See *Note 6*)

1. Liquid handling system.
2. 2 mL 96-deep-well plate.
3. Magnetic stand for 96-deep-well plate.

3 Methods

Carry out all procedures at room temperature unless otherwise specified.

3.1 RNA Extraction

Before starting, mix binding buffer in a rotary shaker.

1. Snap-freeze fresh plant material with liquid nitrogen and grind to a fine powder mechanically using a bead mill homogenizer or manually using a mortar and pestle (*see* **Notes 3** and 7). Pulverized sample can be stored at −80 °C for a long term.
2. Add 200 μL of lysis buffer and immediately suspend frozen samples thoroughly (*see* **Note 8**).
3. Centrifuge at maximum speed for 2 min.
4. Transfer 150 μL of the supernatant to a new 1.5 mL tube.
5. Add 270 μL of SPRI beads and 270 μL of IPA and mix well or vortex for 1 min (*see* **Note 9**).
6. Incubate the homogenates at room temperature for 5 min to bind the RNA to the magnetic beads.
7. Place tubes on a magnetic stand for 3 min.
8. Discard the supernatant and remove tubes from a magnetic stand.
9. Add 300 μL of wash buffer or 85% ethanol and mix well or vortex for 2 min (*see* **Note 10**).
10. Incubate the mixture at room temperature for 1 min.
11. Briefly spin down and place the tube on a magnetic stand for 3 min.
12. Discard the supernatant and remove the tube from a magnetic stand.
13. Add 0.9 mL of 85% ethanol and mix well or vortex for 2 min.
14. Briefly spin down and place the tube on a magnetic stand for 3 min.
15. Discard the supernatant.
16. Repeat **steps 16–18** twice. Remove as much ethanol as possible after the final wash (*see* **Note 11**).
17. Keep the lid open and air-dry at room temperature for 5 min on a magnetic stand (*see* **Note 12**). If performing optional genomic DNA removal, go to Subheading 3.3.
18. Add 100 μL of nuclease-free water and mix well or vortex for 2 min.
19. Incubate at room temperature for 2 min.
20. Place tubes on a magnetic stand for 2 min and transfer 90 μL of the supernatant to a new 1.5 mL tube.

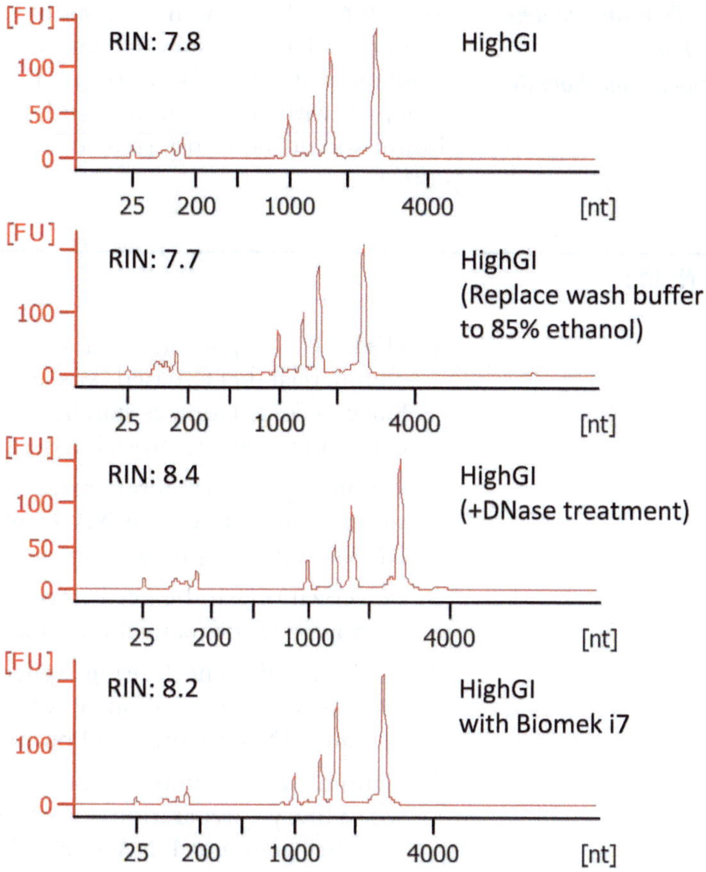

Fig. 1 Electropherograms of total RNA extracted from young shoots using HighGI. The *y*-axis represents fluorescence intensity and the *x*-axis represents the length of the RNA molecule (nucleotides, nt). The RNA integrity number (RIN) values are indicated. The lowest peak at 25 nt is a size marker

21. Measure the RNA concentration and purity with a NanoDrop spectrophotometer. Integrity check by Agilent 2100 Bioanalyzer or 4200 TapeStation is optional (*see* Fig. 1 and **Notes 13** and **14**).
22. Store at −80 °C.

3.2 Genomic DNA Removal (Optional; See *Note 5)*

1. Add 25 μL of DNase I solution (2 U/μL) in 1x DNase I buffer to tubes from **step 20** in Subheading 3.1 and thoroughly resuspend by pipetting.
2. Close the lid and incubate at 37 °C for 15 min.
3. Add 138 μL of wash buffer and mix well or vortex.
4. Incubate tubes at room temperature for 5 min.
5. Briefly spin down and place tubes on DynaMag-2 for 3 min.
6. Discard the supernatant and go to **step 16** in Subheading 3.1.

3.3 Semi-automated RNA Extraction (Optional; See Note 6)

Steps 1–6 in Subheading 3.1 should be performed manually, and transfer 150 μL of supernatant to a 2 mL 96-deep-well plate. Subsequent steps can be automated using a liquid handling system equipped with a magnetic stand for 96-deep-well plate. Final elution should be transferred to a 96-well PCR plate and kept at −80 °C.

4 Notes

1. HighGI is applicable to a variety of tissue types for which commercial GTC-based kits are used for RNA extraction. Since GTC gelatinizes starch, HighGI is not suitable for RNA extraction from starch-rich tissues such as endosperm.
2. We anticipate that input materials can be reduced. We have successfully extracted RNA from two root tips enough for RNA-seq library preparation.
3. Pulverization can be performed mechanically with a bead mill homogenizer or manually with a mortar and pestle.
4. RNA integrity check using Agilent Bioanalyzer is optional. If there is no concern about RNA degradation, it is sufficient to measure RNA quantity and purity using NanoDrop.
5. Genomic DNA removal is not necessarily required for mRNA-seq library preparation or Northern blotting. However, optional genomic DNA removal is highly recommended for genomic DNA sensitivity assays, such as reverse transcription quantitative PCR.
6. HighGI can be semi-automated using commercial liquid handling systems such as Beckman Biomek i7. Buffers used in HighGI do not contain hazardous substances such as phenol or chloroform, so all necessary reagents can be safely dispensed in advance in a liquid handling system.
7. If using a bead mill homogenizer and placing leaves into the tube, fold them into a U shape to prevent the leaves from sticking to the side of the tube during pulverization.
8. A rotary shaker can be used to expedite this process.
9. Binding buffer is highly viscous and should be pipetted slowly.
10. We have not observed any difference in yield and integrity of RNA when wash buffer or 85% ethanol was used in this step. But it is expected that the use of wash buffer would effectively inhibit residual RNase activity.
11. Residual GTC in ethanol significantly lowers the 260/230 absorbance ratio but has little effect on most downstream assays.

12. Avoid overdrying. It may result in a lower recovery.
13. RNA extracted by HighGI method retains low molecular weight RNAs less than 200 bp and can be used for small RNA analysis, including microRNAs or small interfering RNAs (*see* Fig. 1).
14. Agilent Bioanalyzer provides RNA integrity number (RIN) values ranging from 1 to 10 as an indicator of RNA integrity [13]. Lower RIN values indicate that RNA is more degraded. RIN values are optimized for mammalian RNA and highly depend on the peak patterns around 17S and 28S ribosomal RNAs (rRNAs). Plant green tissue is rich in chloroplasts, whose rRNA peaks are located around 17S and 28S rRNAs, resulting in lower RIN values in plant green tissue.

Acknowledgments

This work was supported by JST CREST (grant number JPMJCR17O1) and JSPS KAKENHI (grant numbers 23H04965 and 23H04968).

References

1. MacDonald RJ, Przybyla AE, Rutter WJ (1977) Isolation and in vitro translation of the messenger RNA coding for pancreatic amylase. J Biol Chem 252:5522–5528
2. Chomczynski P, Sacchi N (1987) Single-step method of RNA isolation by acid guanidinium thiocyanate-phenol-chloroform extraction. Anal Biochem 162:156–159
3. Logemann J, Schell J, Willmitzer L (1987) Improved method for the isolation of RNA from plant tissues. Anal Biochem 163:16–20
4. Chang S, Puryear J, Cairney J (1993) A simple and efficient method for isolating RNA from pine trees. Plant Mol Biol Report 11:113–116
5. Jaakola L, Pirttilä AM, Halonen M et al (2001) Isolation of high quality RNA from bilberry (Vaccinium myrtillus L.) fruit. Mol Biotechnol 19:201–203
6. Zeng Y, Yang T (2002) RNA isolation from highly viscous samples rich in polyphenols and polysaccharides. Plant Mol Biol Report 20: 417a–417e
7. Barbier FF, Chabikwa TG, Ahsan MU et al (2019) A phenol/chloroform-free method to extract nucleic acids from recalcitrant, woody tropical species for gene expression and sequencing. Plant Methods 15:62
8. Arakawa T, Timasheff SN (1984) Protein stabilization and destabilization by guanidinium salts. Biochemistry 23:5924–5929
9. Ullrich A, Shine J, Chirgwin J et al (1977) Rat insulin genes: construction of plasmids containing the coding sequences. Science 196: 1313–1319
10. Chirgwin JM, Przybyla AE, MacDonald RJ et al (1979) Isolation of biologically active ribonucleic acid from sources enriched in ribonuclease. Biochemistry 18:5294–5299
11. Rohland N, Reich D (2012) Cost-effective, high-throughput DNA sequencing libraries for multiplexed target capture. Genome Res 22:939–946
12. Yoshino K, Nishijima R, Kawakatsu T (2020) Low-cost RNA extraction method for highly scalable transcriptome studies. Breed Sci 70: 481–486
13. Schroeder A, Mueller O, Stocker S et al (2006) The RIN: an RNA integrity number for assigning integrity values to RNA measurements. BMC Mol Biol 7:3

Chapter 3

Total RNA Extraction from Starchy Rice Endosperm

Yuhya Wakasa

Abstract

RNA extraction is a critical step in gene expression analysis. While numerous RNA extraction kits are commercially available, most kits cannot be utilized for RNA extraction from rice endosperm that contains abundant starch. Here, I describe a three-step RNA extraction from rice starchy endosperm. The process involves phenol/chloroform extraction, ethanol precipitation, and lithium chloride precipitation. This method yields an abundant and high-quality total RNA even from a single starchy endosperm.

Key words Phenol/chloroform, Lithium chloride precipitation, Rice seed, Starchy endosperm, Total RNA extraction

1 Introduction

It is believed that when life arose on Earth, it was initially "an RNA world" in which the main catalyst of interactions between cells was RNA [1, 2]. Later, when life on Earth transitioned to "the DNA world," RNA molecules continued to be a central actor in the gene expression and protein synthesis of organisms. Since the dawn of molecular biology, the behavior of RNA molecules has been a central and enduring topic for scientific researchers, including those who study rice (*Oryza sativa* L.). RNA extraction is the first basic step in RNA research. In recent years, many kinds of kits suitable for RNA extraction have been developed, and there is little problem in extracting RNA from most samples. However, in the case of the starchy endosperm in rice, even sophisticated kits may not be able to successfully extract the total RNA, such as providing insufficient quantity. Thus, what should we do? The conventional method using lithium chloride precipitation would be quite suitable for extracting RNA from the starchy endosperm of rice [3] and is satisfactory enough for practical use. This method consists of three simple steps: (a) phenol/chloroform extraction, (b) ethanol precipitation, and (c) LiCl precipitation. RNA extracted by this

Taiji Kawakatsu (ed.), *Rice: Methods and Protocols*, Methods in Molecular Biology, vol. 2869,
https://doi.org/10.1007/978-1-0716-4204-7_3,

method can be used in a variety of experiments, such as Northern blot, RT-PCR, quantitative RT-PCR, and RNA-seq [4–7].

2 Materials

All labware and reagents should be RNase-free.

2.1 RNA Extraction

1. Liquid nitrogen.
2. Bead mill homogenizer or mortar and pestle.
3. RNase-free water or diethyl pyrocarbonate (DEPC)-treated water (*see* **Note 1**).
4. A mixture of acid phenol/chloroform/isoamyl alcohol (25: 24:1) (*see* **Note 2**).
5. RNA extraction buffer: 0.1 M Tris–HCl, pH 9.0, 1% sodium dodecyl sulfate (SDS), 0.1 M NaCl, and 5 mM EDTA. Autoclave at 121 °C for 40 min and store at room temperature.
6. 3 M sodium acetate, pH 5.2.
7. 99% ethanol.
8. 70% ethanol.
9. 8 M lithium chloride.
10. 2 M lithium chloride.
11. UV-Vis spectrophotometer.
12. Agarose gel electrophoresis device.
13. 50× TAE: 2 M Tris base, 1 M acetic acid, 50 mM EDTA, pH 8.0.
14. 1% agarose gel in 1× TAE.

2.2 DNase I Treatment

1. RNase-free water or DEPC-treated water (*see* **Note 1**).
2. DNase I (5 units/μL).
3. 10× DNase I reaction buffer: 100 mM Tris–HCl, 25 mM $MgCl_2$, 5 mM $CaCl_2$, pH 7.6.
4. RNase inhibitor (40 units/μL).
5. A mixture of acid phenol/chloroform/isoamyl alcohol (25: 24:1) (*see* **Note 2**).
6. 3 M sodium acetate, pH 5.2.
7. 99% ethanol.
8. 70% ethanol.

3 Methods

This protocol is for "small scale" RNA extraction from one to four starchy endosperms in a 1.5–2.0 mL tube, but can be scaled up, as necessary. In this chapter, the term "starchy endosperm" means seed tissues from 1 to 35 days after flowering. After the total RNA extraction, perform DNase I treatment, if necessary.

3.1 RNA Extraction

1. Harvest one to four starchy endosperms from a rice plant into the tube and immediately put the entire tube into liquid nitrogen. If necessary, remove the lemma and the palea from seeds smoothly before putting into liquid nitrogen.
2. Grind the starchy endosperm into a fine powder with a bead mill homogenizer or with a pestle and mortar. If the latter, collect the powder in a 1.5 mL or 2.0 mL tube quickly before it thaws.
3. Add 400 μL of the acid phenol/chloroform/isoamyl alcohol into the tube and mix quickly.
4. Add 400 μL of RNA extraction buffer into the tube and mix well by vortexing for more than 15 min.
5. Centrifuge at 18000–20000× *g* at room temperature for 10 min.
6. Collect the supernatant into a new tube.
7. Add 400 μL of acid phenol/chloroform/isoamyl alcohol into the tube and mix well by vortexing for several seconds.
8. Centrifuge at 18000–20000× *g* at room temperature for 10 min.
9. Collect the supernatant into a new tube.
10. Add 40 μL of 3 M sodium acetate and 1 mL of 99% ethanol into the tube and mix well by turning upside down several times.
11. Place on ice for more than 10 min.
12. Centrifuge at 18000–20000× *g* at 4 °C for 10 min.
13. Discard the supernatant.
14. Wash the pellet with 1 mL of 70% ethanol.
15. Dry the pellet briefly (*see* **Note 3**).
16. Dissolve the pellet in 60 μL of RNase-free water.
17. Add 20 μL of 8 M lithium chloride into the tube and mix well by pipetting or tapping.
18. Place on ice for more than 1 h. If the amount of RNA is expected to be low, it may be placed overnight in a cold room at 4 °C.

19. Centrifuge at 18000–20000× *g* at 4 °C for 10 min.
20. Discard the supernatant.
21. Wash the pellet with 1 mL of 2 M lithium chloride.
22. Wash the pellet well with 70% ethanol (*see* **Note 4**).
23. Dry the pellet briefly (*see* **Note 3**).
24. Dissolve the pellet in 50–100 μL of RNase-free water. Up to several hundred micrograms of total RNA would usually be obtained.
25. Estimate the total RNA yield and quality by using an UV-Vis spectrophotometer and agarose electrophoresis (*see* **Note 5**).

3.2 DNase I Treatment

1. Prepare the DNase I mixture in a 1.5 mL tube as follows:

Total RNA from **step 25** in Subheading 3.1	Up to 100 μg
10× DNase I reaction buffer	10 μL
DNase I	1 μL
RNase inhibitor	1 μL
RNase-free water	Up to 100 μL
Total	100 μL

2. Incubate at 37 °C for 1 h.
3. Add 100 μL of acid phenol/chloroform/isoamyl alcohol into the tube and mix well by vortexing for several seconds.
4. Transfer the supernatant into a new tube.
5. Add 10 μL of sodium acetate and 250 μL of 99% ethanol and mix well by turning upside down several times.
6. Place on ice for 10 min.
7. Centrifuge at 18000–20000× *g* at 4 °C for 10 min.
8. Discard the supernatant.
9. Wash the pellet with 1 mL of 70% ethanol.
10. Dry the pellet briefly (*see* **Note 3**).
11. Dissolve the pellet in the desired amount of RNase-free water.

4 Notes

1. Autoclaving may not alone fully inactivate RNase in solutions. RNase-free reagents can be purchased from various suppliers. Or prepare RNase-free reagents by DEPC treatment. For

DEPC treatment, add 1 mL of DEPC to 1 L of distilled water or the reagent of choice. DEPC separates from the aqueous solution, similar to oil. Shake vigorously until DEPC is completely dissolved in the solution. Autoclave at 121 °C for 15 min. While DEPC is suspected as a carcinogen and should be handled with care, autoclaving inactivates DEPC. Note that a residual smell may persist after autoclaving. DEPC treatment is not effective for Tris or HEPES buffer.

2. Acid phenol/chloroform/isoamyl alcohol mixture is highly preferred compared to TE-saturated phenol/chloroform/isoamyl alcohol. The acidic environment ensures that DNA partitions into organic phase, while RNA remains in the aqueous phase. On the other hand, at a neutral or alkaline condition, both DNA and RNA remain in the aqueous phase. Acid phenol/chloroform/isoamyl alcohol is commercially available from various suppliers or can be prepared as follows. Thaw 500 g of crystalline phenol in the supplied bottle by incubating in a water bath at 60–70 °C until liquid. Add 150–200 mL water and 0.5 g of 8-hydroxyquinoline and mix well with a stirrer and a magnetic stirrer bar. Place in a cold room until the upper aqueous layer and the lower phenol layer separate. This reagent is called acidic phenol or water-saturated phenol. Remove the excess aqueous layer with an aspirator. Add 240 mL of chloroform and 10 mL of isoamyl alcohol into a 500 mL graduated cylinder. Make up to a volume of 500 mL with acidic phenol. Note that addition of 250 mL of acidic phenol does not make 500 mL of phenol/chloroform/isoamyl alcohol, because the acidic phenol is dehydrated when mixed with chloroform. Therefore, the amount required to make 500 mL will be more. Mix well and place in a cold room until the upper aqueous layer and the lower organic layer separate. Remove the excess aqueous layer (top) with an aspirator. Store at 4 °C in a cold room.

3. This wash step with 70% ethanol is critical because the residual lithium chloride (lithium ion) inhibits reverse transcriptase activity.

4. Avoid overdrying, because overdried RNA is difficult to dissolve in water.

5. Even a minimal contamination of RNase I in the agarose gel or electrophoresis buffer can lead to RNA degradation during electrophoresis. Concerns regarding RNA integrity may warrant the use of denaturing formaldehyde agarose gel, which can provide a more protective environment for RNA molecules during the electrophoresis process [8].

References

1. Gilbert W (1986) Origin of life: the RNA world. Nature 319:618–618
2. Higgs PG, Lehman N (2015) The RNA world: molecular cooperation at the origins of life. Nat Rev Genet 16:7–17
3. Takaiwa F, Kikuchi S, Oono K (1987) A rice glutelin gene family: a major type of glutelin messenger-RNAs can be divided into 2 classes. Mol Gen Genomics 208:15–22
4. Tada Y, Utsumi S, Takaiwa F (2001) Foreign gene products can be enhanced by introduction into low storage protein mutants. Plant Biotechnol J 1:411–422
5. Kawakatsu T, Yamamoto MP, Hirose S et al (2008) Characterization of a new rice glutelin gene GluD-1 expressed in the starchy endosperm. J Exp Bot 59:4233–4245
6. Wakasa Y, Yasuda H, Oono Y et al (2011) Expression of ER quality control-related genes in response to changes in BiP1 levels in developing rice endosperm. Plant J 65:675–689
7. Kawakatsu T, Kawahara Y, Itoh T et al (2013) A whole-genome analysis of a transgenic rice seed-based edible vaccine against cedar pollen allergy. DNA Res 20:623–631
8. Sambrook J, Russell DW (2001) Protocol 6. Separation of RNA according to size: electrophoresis of RNA through agarose gels containing formaldehyde. In: Molecular cloning: a laboratory manual, vol 1, 3rd edn. Cold Spring Harbor Laboratory Press, New York, pp 7.31–7.34

Chapter 4

RNA Extraction from Rice Immature Embryo Using Laser Capture Microdissection

Takumi Tezuka, Misuzu Nosaka-Takahashi, Takanori Yoshikawa, and Yutaka Sato

Abstract

Laser capture microdissection (LCM) enables the selective isolation of organs, tissues, and cells from surrounding tissues. Total RNA extracted from small tissue sections can be used for a variety of subsequent analysis such as RNA-seq analysis. Here, we describe a method for isolating embryos from rice ovary sections using LCM and extracting total RNA. We successfully obtained sufficient amount of total RNA from a single rice embryo at the globular embryo stage for RNA-seq analysis.

Key words Laser capture microdissection (LCM), Paraffin section, RNA extraction, RNA-seq, Embryo

1 Introduction

Laser capture microdissection (LCM) allows for the precise isolation of specific organs, tissues, and cells from surrounding tissues within the section by laser under a microscope after fixing and sectioning the samples. Initially developed for collecting cancer cells in mammals [1, 2], LCM techniques have since been adapted for tissue isolation in both mammals and plants [3–5]. During this process, the composition of fixatives and sectioning conditions for various plant tissues have been optimized [5, 6]. Additionally, slides equipped with transparent transfer films (thermoplastic polymer films) have been employed for tissue sectioning [4, 5]. When exposed to the laser, this film fuses with the target tissue, which can be collected in a tube without destruction of cell or tissue structure. Methodological advancements have significantly reduced RNA degradation during LCM procedures.

Because plant embryos are small and difficult to isolate, LCM is a powerful tool for the study of embryogenesis. LCM enables the precise isolation of whole embryo under the microscope, reducing

Taiji Kawakatsu (ed.), *Rice: Methods and Protocols*, Methods in Molecular Biology, vol. 2869,
https://doi.org/10.1007/978-1-0716-4204-7_4,

contamination from surrounding tissues [6, 7]. Moreover, by isolating specific regions of the rice embryo, region-specific expression genes were discerned [8]. Here, we describe a method for isolating a single embryo at the globular embryo stage from a section of a rice ovary using LCM, followed by total RNA extraction from the section. With this approach, the transcriptomes of wild-type and mutant embryos can be compared, allowing for the identification of differentially expressed genes that characterize the mutant across the genome [9, 10].

2 Materials

All solutions should be prepared using RNase-free sterile water to prevent RNA degradation. Fixative and ethanol solutions should be stored below 4 °C until dehydrating the sample with 100% ethanol. It is imperative to diligently adhere to all waste disposal regulations when disposing of waste materials.

2.1 Sample Fixation

1. Fixative solution: 3:1 ethanol/acetate. Fixative is prepared just before using. Store at 4 °C.
2. Aspirator.
3. Desiccator.

2.2 Paraffin Embedding Using Microwave Processor

1. Microwave processor H2580 (Energy Beam Sciences).
2. Graded ethanol series: 80%, 90%, and 100% ethanol. Dilute 100% ethanol with sterile distilled water.
3. 1:1 ethanol/n-butanol.
4. Paraffin wax: Melt Paraplast Xtra (Leica Biosystems) by incubating at 60 °C.
5. 1:1 n-butanol/paraffin wax: After mixing, tighten the cap and store at 60 °C in paraffin oven.
6. Hot plate for paraffin embedding.

2.3 Sectioning

1. RNA secure solution: Dilute RNA secure RNase Inactivation Reagent (Invitrogen) to a 1:25 ratio with sterile water.
2. Rotary microtome.
3. Polyethylene naphthalate (PEN) membrane slide, 2 μm (Leica Microsystems).
4. Plastic Coplin jar.
5. Histo-Clear II.

2.4 LCM and RNA Extraction

1. LMD7000 (Leica Microsystems).
2. Collection tube for LCM (*see* **Note 1**).
3. PicoPure RNA isolation kit (Invitrogen).
4. 2100 Bioanalyzer Instrument (Agilent).
5. RNA 6000 Pico kit (Agilent).

3 Methods

Perform all experiments carefully to avoid RNase contamination.

3.1 Sample Fixation

1. Peel off palea and lemma from the rice spikelet using tweezers or razors and collect the carpel. Trim the apical tip of the carpel to facilitate easy penetration of the fixative solution.
2. Immerse approximately 20 carpels into 10 mL of pre-chilled fixative solution. While sampling, the vial of fixative solution should be placed on ice.
3. Put the vial in a desiccator and vacuum the vial using aspirator for 30 min (*see* **Note 2**).
4. Release the vacuum slightly until pressure is restored.
5. Replace the fixative solution with fresh solution.
6. Repeat **steps 3–5** three times. The vial of fixative solution should be placed on ice throughout these procedures. After the fixation, samples can be stored in the fixative solution at 4 ° C for 1–3 days.

3.2 Paraffin Embedding Using Microwave Processor

1. Place the vial containing the sample on ice and expose to microwave at 250 W for 15 min using a microwave processor. Ensure that the temperature of the fixative solution does not exceed 20 °C (*see* **Note 3**).
2. Replace the fixative solution with fresh solution.
3. Repeat **steps 1–2** twice.
4. Dehydrate the sample by replacing the fixative solution with a graded ethanol series (increasing concentrations, 80%, 90%, and 100%). Replace the solution and expose to microwave at 300 W for 1.5 min in each concentration. From this step onward, the sample's temperature is gradually increased until it reaches a final temperature of 58 °C.
5. Repeat **step 4** with fresh 100% ethanol.
6. Clear the sample by replacing ethanol with n-butanol. Replace the solution with 1:1 ethanol/n-butanol and expose to microwave at 300 W for 1.5 min.
7. Repeat **step 6** with 100% n-butanol.

8. Infiltrate the samples with paraffin wax. Replace the solution with 1:1 n-butanol/paraffin wax and microwave at 300 W for 10 min (*see* **Note 4**).
9. Replace the solution with paraffin wax and microwave at 300 W for 10 min.
10. Replace the solution with fresh paraffin wax and microwave at 250 W for 30 min.
11. Repeat **step 10** three times.
12. Prepare a plastic dish containing paraffin wax and place it on hot plate at 60 °C.
13. Transfer the content of the vial, including samples, into the plastic dish.
14. Transfer the samples using tweezers to a plastic dish prepared at **step 12**.
15. After cooling, store the paraffin blocks at 4 °C.

3.3 Sectioning

1. Incubate dilute RNA secure solution at 60 °C.
2. Dispense 400 μL of RNA secure solution onto a PEN membrane slide pre-warmed to 57 °C on a hot plate.
3. Cut 14-μm-thick serial sections from paraffin blocks using a rotary microtome.
4. Place the serial sections on the RNA secure solution from **step 2.** Allow approximately 30 s for the sections to stretch. Once stretched, promptly remove the solution.
5. Place slides in a plastic Coplin jar at 4 °C for 1–2 h. The slide produced should be used for LCM the same day.
6. To remove paraffin wax, immerse the slides in Histo-Clear II for 10 min.
7. Repeat **step 6** with fresh Histo-Clear II.
8. Place the slides at room temperature to dry for 30 min.

3.4 LCM and RNA Extraction

1. Collect the selected areas in the sections by LCM (*see* Fig. 1 and **Note 5**). The target tissues are collected in the lid of the collection tube. Approximately five serial sections, each 14 μm thick, collectively encompass a single rice embryo at the globular stage. When the entire single embryo is retrieved from these serial sections by LCM, the total area of the sections is approximately 0.1 mm^2.
2. Add 50 μL of extraction buffer in PicoPure RNA isolation kit to the lid of the collection tube containing the target tissue.
3. Carefully close the lid of the collection tube and centrifuge at 800× *g* for 2 min to collect the liquid at the bottom of the tube.

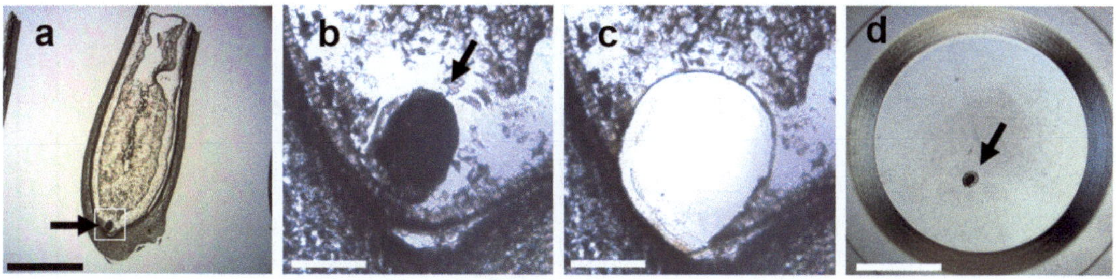

Fig. 1 Isolation of globular stage embryos from rice carpel by LCM. (**a**) Whole rice carpel section on 3 days after pollination. (**b**) Enlarged view of the area within the white box in (**a**). (**c**) The same region as in (**b**), following the LCM dissection of the embryo. (**d**) A single section depicting part of an embryo, captured on the lid of the collection tube. Arrows indicate rice embryos. Bars = 1 mm (**a**, **d**) and 100 μm (**b**, **c**)

4. Incubate the tube at 42 °C for 30 min to extract the RNA from the sections. Samples can be stored at −80 °C.
5. Extract total RNA from LCM samples using PicoPure RNA isolation kit, according to the manufacturer's protocol. Perform genomic DNA degradation by DNase as described in the appendix of the protocol. Elute RNA with 20 μL of elution buffer.
6. Assess the quality of the extracted RNA using the 2100 Bioanalyzer and the RNA 6000 Pico kit, according to manufacturer's protocol (*see* **Note 6**).

4 Notes

1. Collection tubes are specified by the LCM system. We use 0.2 mL PCR tubes (Corning).
2. This step ensures fixation of the samples. We utilize an aspirator to decompress, gradually reducing the pressure inside the desiccator to −0.1 MPa over 10 min or longer.
3. Irradiate the sample with microwaves to promote displacement of the fixative solution. Ensure that the temperature of the fixative solution remains at 20 °C or lower, as elevated temperatures may lead to RNA degradation. Place the vial on ice within the microwave processor, and add fresh ice as needed to prevent melting and maintain the low temperature of the fixative solution.
4. If the temperature is too low and the paraffin solidifies, repeat microwaving under the same conditions.
5. LCM systems should be operated in accordance with manufacturer's protocol. Given that laser conditions can vary among the kind of tissues, it is essential to fine-tune the optimal settings for each tissue. The stronger the laser power, the larger

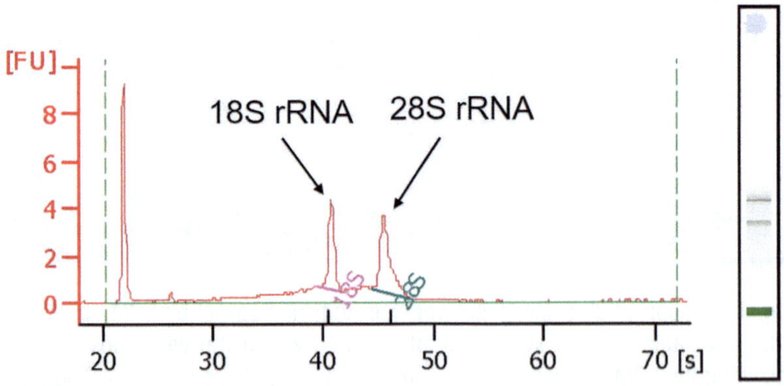

Fig. 2 Electropherogram of total RNA extracted from five sections of one rice embryo at the globular stage isolated by LCM. If the RNA is not degraded, the 18S and 28S rRNA peaks can be identified

area is ablated. To achieve a more precise cut of the section, it is essential to determine the minimum laser power required to cut the section accurately. We use the LMD7000 under laser conditions of power 29, aperture 10, and speed 9 for embryos.

6. The RNA 6000 Pico kit requires total RNA within the range of 200–5000 pg/μL for accurate measurement. RNA from one embryo typically falls within this range, but if it contains a higher concentration of RNA, it should be diluted before measurement. Distinct peaks corresponding to 18S and 28S rRNA suggest that the RNA was extracted intact, without degradation (*see* Fig. 2). RNA quality is also determined by the RNA integrity number (RIN); the RIN value of 5 or higher is recommended for cDNA synthesis. The result shown in Fig. 2 is the total RNA extracted from five sections of one rice embryo at the globular stage. Bioanalyzer measurement shows RIN of 8.2 and RNA concentration of 324 pg/μL. When performing RNA-seq on LCM samples, we utilize SMART-Seq HT Kit (Takara) to synthesize cDNA from mRNA. The kit is suitable for RNA-seq from a small amount of total RNA as PCR amplification is performed during cDNA synthesis.

References

1. Emmert-Buck MR, Bonner RF, Smith PD et al (1996) Laser capture microdissection. Science 274:998–1001
2. Bonner RF, Emmert-Buck M, Cole K et al (1997) Laser capture microdissection: molecular analysis of tissue. Science 278:1481–1483
3. Asano T, Masumura T, Kusano H et al (2002) Construction of a specialized cDNA library from plant cells isolated by laser capture microdissection: toward comprehensive analysis of the genes expressed in the rice phloem. Plant J 32:401–408
4. Nakazono M, Qiu F, Borsuk LA et al (2003) Laser-capture microdissection, a tool for the global analysis of gene expression in specific plant cell types: identification of genes

expressed differentially in epidermal cells or vascular tissues of maize. Plant Cell 15:1049

5. Kerk NM, Ceserani T, Tausta LS et al (2003) Laser capture microdissection of cells from plant tissues. Plant Physiol 132:27–35
6. Takahashi H, Kamakura H, Sato Y et al (2010) A method for obtaining high quality RNA from paraffin sections of plant tissues by laser microdissection. J Plant Res 123:807–813
7. Chen M, Lin JY, Wu X et al (2021) Comparative analysis of embryo proper and suspensor transcriptomes in plant embryos with different morphologies. Proc Natl Acad Sci USA 118(6):e2024704118
8. Itoh JI, Sato Y, Sato Y et al (2016) Genome-wide analysis of spatiotemporal gene expression patterns during early embryogenesis in rice. Development 143:1217–1227
9. Ishimoto K, Sohonahra S, Kishi-Kaboshi M et al (2019) Specification of basal region identity after asymmetric zygotic division requires mitogen-activated protein kinase 6 in rice. Development 146:dev176305
10. Ta KN, Yoshida MW, Tezuka T et al (2023) Control of plant cell growth and proliferation by MO25A, a conserved major component of the Mammalian Sterile 20-like kinase pathway. Plant Cell Physiol 64:336–351

Chapter 5

Using Immunoblotting for Detection of a Circadian Clock Component, Os–GI, in Rice Leaves

Hironori Itoh

Abstract

The adaptability of plants to given environmental conditions is achieved through the regulation of various genes. The functional regulation of genes depends on precise transcriptional and translational controls that, at the end, determine the expression of the translated products (proteins). Here, I describe the immunoblotting method for the detection of proteins expressed in rice leaves. This method includes protein extraction, protein separation, electroblotting, and immune detection.

Key words Protein quantification, SDS–PAGE, Electroblotting, Immunoblotting

1 Introduction

Plants can recognize fluctuations in living environments and, in response, synchronize their growth and differentiation. They recognize seasonal changes, such as day length and temperature, as external signals and perceive biotic and abiotic stresses. Environmental responses are thought to be efficiently adapted to both long- and short-term environmental changes by the appropriate transcriptional and post-translational regulation of the corresponding genes and proteins. At the post-transcriptional level, proteins may be regulated through a diversity of mechanisms, such as protein stability, modification, and subcellular localization. Noteworthy, protein analyses may help in elucidating the detailed molecular function of a gene of interest. Key analysis methods include protein quantification in prepared crude extracts, protein separation (based on their molecular masses) by SDS–PAGE, and detection of target proteins using epitope peptide–specific monoclonal antibodies.

In this chapter, I analyzed the expression of GIGANTEA (Os–GI), in rice plants to exemplify the use of the methods mentioned above. Given that GI is a component of plant circadian

Taiji Kawakatsu (ed.), *Rice: Methods and Protocols*, Methods in Molecular Biology, vol. 2869,
https://doi.org/10.1007/978-1-0716-4204-7_5,

clocks, GI protein levels are diurnally regulated [1]. It has been hypothesized that GI levels are also regulated by proteasome-dependent degradation [2] (*see* **Note 1**).

2 Materials

1. Extraction buffer: 20 mM Tris–HCl, pH 7.5, 150 mM NaCl, 0.5% Tween 20, 1 mM EDTA, 9–14 units DNase I, 10 μg RNase A, 1× cOmplete Mini EDTA-free protease inhibitor (*see* **Note 2**).
2. Bio-Rad Protein Assay Dye reagent concentrate.
3. Bovine serum albumin (BSA) dilution series: Dilute the stock BSA solution (20 mg/mL) to 0.5, 1, 2, 3, 4, 5, and 10 μg/μL with extraction buffer without DNase, RNase, and protease inhibitor.
4. SDS sample buffer (2×): 135 mM Tris–HCl, pH 6.8, 200 mM dithiothreitol, 4% sodium dodecyl sulfate (SDS), 0.2% bromophenol blue, 20% glycerol. Prepare 10 mL. Store aliquots at −30 °C for long-term use.
5. 30% acrylamide/bis-acrylamide solution: 29% (w/v) acrylamide, 1% (w/v) bis-acrylamide.
6. Separation gel buffer: 1.5 M Tris–HCl, pH 8.8.
7. Stacking gel buffer: 1 M Tris–HCl, pH 6.8.
8. 20% SDS.
9. 10% ammonium persulfate (APS): Store at 4 °C up to 1 week.
10. N,N,N′,N′-tetramethylethylenediamine (TEMED).
11. Glass plates, spacers, comb, and binder clips for gel preparation.
12. Isopropyl alcohol.
13. Tris–glycine electrophoresis buffer (5×): 125 mM Tris, 960 mM glycine, 0.5% SDS. Dilute to 1× with sterile water before use.
14. Molecular weight protein marker.
15. Syringe with 21 G needle.
16. Filter paper: Whatman cellulose chromatography paper 3 MM Chr.
17. PVDF membrane, pore size 0.45 μm.
18. Wet/tank-type electroblotting apparatus.
19. Electroblotting buffer: 25 mM Tris, 192 mM glycine, 20% (v/v) methanol. Store at 4 °C.

20. Ponceau S solution: Add 0.1 g Ponceau S to a 200 mL beaker containing 190 mL sterile water. Add 5 mL acetic acid, and mix. Make up to 200 mL with sterile water. Store at room temperature. This solution can be reused several times in the short term.
21. 10× phosphate-buffered saline (PBS): 1.37 M NaCl, 27 mM KCl, 81 mM Na_2HPO_4, 14.7 mM KH_2PO_4. Adjust pH to 7.4 and autoclave.
22. PBS-T: 1× PBS containing 0.1% Tween-20.
23. Blocking buffer: 5% non-fat skimmed milk in PBS-T (*see* **Note 3**).
24. Primary antibody: Anti-HA-tag mouse monoclonal antibody.
25. Secondary antibody conjugated with horseradish peroxidase (HRP): Anti-mouse IgG antibody conjugated with HRP.
26. Chemiluminescence reagent.

3 Methods

3.1 Preparation of Crude Extracts from Rice Leaf Samples

1. Collect leaf samples, weight them, and immediately freeze them in liquid nitrogen (*see* **Note 4**). Store at −80 °C.
2. Add liquid nitrogen to both a mortar and pestle and wait until they are completely chilled. Transfer the samples into the liquid nitrogen-chilled mortar and grind them into a fine powder.
3. Add twice the volume (v/w) of extraction buffer.
4. Firmly grind the samples with the pestle until they can be collected with a large orifice pipette tip.
5. Transfer all the sample, including all the solution and debris, to a chilled 1.5 mL tube.
6. Centrifuge at 19,000× *g* for 15 min at 4 °C.
7. Transfer the supernatant to a new 1.5 mL tube.
8. Repeat **step 6** and transfer the clear supernatant, i.e., the crude protein extract, to a new 1.5 mL tube.

3.2 Protein Quantification in the Crude Extract

Protein concentration can be quantified by colorimetric methods [3]. Protein concentration in the samples can be determined by comparison with a standard curve; in this case, I used a BSA curve. Crude protein extracts may be diluted with extraction buffer if needed (*see* **Note 5**).

1. Transfer 1 μL of extraction buffer (as blank), each solution in the BSA dilution series (corresponding to 0.5, 1, 2, 3, 4, 5, 7.5, and 10 μg; standard), and crude extracts (unknown samples) from **step 8** in Subheading 3.1 to individual 1.5 mL tubes (*see* **Note 5**).

2. Add 800 μL of sterile water and 200 μL of Protein Assay Dye reagent concentrate to each tube. Gently mix by inverting the tubes.
3. Incubate at room temperature for 10 min.
4. Measure the absorbance of each of the solutions at 595 nm (A595) using a spectrophotometer.
5. Plot the A595 values of the BSA dilution series solutions versus protein concentrations to prepare a standard calibration curve. Based on this calibration curve, the protein concentration in a sample can be determined from its A595 value.

3.3 Sample Preparation for SDS–PAGE

1. Mix 30 μg of total protein (the volume will vary depending on the protein concentration in each sample solution) and 10 μL 2× SDS sample buffer in a 1.5 mL tube (*see* **Note 6**). Add sterile water to 20 μL. Mix by finger tapping and flash spin all samples in a microcentrifuge.
2. Incubate at 95 °C for 5 min.
3. Flash spin all samples in a microcentrifuge.

3.4 Protein Separation by SDS–PAGE

1. Cast the two glass plates appropriately, with the aid of the spacers and the binder clips, to make the SDS–PAGE gel.
2. Prepare the separation gel solution. Mix the following components (*see* **Note 7**):

Sterile water	4.7 mL
30% acrylamide/bis-acrylamide solution	2.65 mL (final 8%)
1.5 M Tris–HCl (pH 8.8)	2.5 mL
20% SDS	0.05 mL
Total	9.9 mL

3. Add 0.1 mL 10% APS and 6 μL TEMED and mix gently. Avoid bubble formation.
4. Pour 6.5 mL of this separation gel solution into the casted glass plates. Gently overlay 1 mL of sterile water onto the top of the separation gel solution for sealing.
5. Let the separation gel polymerize at room temperature for 30–60 min.
6. Decant and discard the overlaid sterile water. Rinse with fresh sterile water to remove any unpolymerized separation gel solution. Tilt the gel 90 degrees and remove excess water with a filter paper without touching the gel.

7. Prepare the stacking gel solution by mixing the following components:

Sterile water	6.9 mL
30% acrylamide/bis-acrylamide solution	1.7 mL
1 M Tris–HCl (pH 6.8)	1.25 mL
20% SDS	0.05 mL
Total volume	9.9 mL

8. Add 0.1 mL 10% APS and 10 μL of TEMED and mix gently, and avoid bubble formation.
9. Remove the filter paper that was inserted into the casted gel plate from **step 6** and pour the stacking gel solution over the polymerized separation gel.
10. Insert a comb and let the separation gel polymerize at room temperature for 30–60 min. While inserting the comb, avoid bubble formation.
11. Assemble the electrophoresis tank and mount the gel inside.
12. Fill the electrophoresis tank with Tris–glycine electrophoresis buffer (1×) and then remove the comb from the gel. Soon after removing the comb, wash any unpolymerized stacking gel solution from the wells with Tris–glycine electrophoresis buffer (1×) using a syringe with a 21 G needle. Completely remove bubbles from the bottom of the gel wells.
13. Load the molecular weight protein marker and the heat-denatured protein samples from in Subheading 3.3 into the gel wells.
14. Run the electrophoresis at a constant current of 20 mA (*see* **Note 8**).
15. While running gel electrophoresis, prepare two sets of three pieces of filter paper and one piece of PVDF transfer membrane for electroblotting. Cut the PVDF transfer membrane about 0.5 cm larger than the gel and the filter paper pieces about 0.5 cm larger than the PVDF membranes. Soak PVDF membranes in methanol for 1 min. Immerse filter papers and pre-wet PVDF membranes into electroblotting buffer.
16. Stop the electrophoresis process when the target protein size ranges indicated by the molecular marker are properly separated.

3.5 Electroblotting

1. Remove the gel from the glass plates and immerse it into electroblotting buffer. To avoid gel breakage, shake it slowly with a shaker for 15 min.

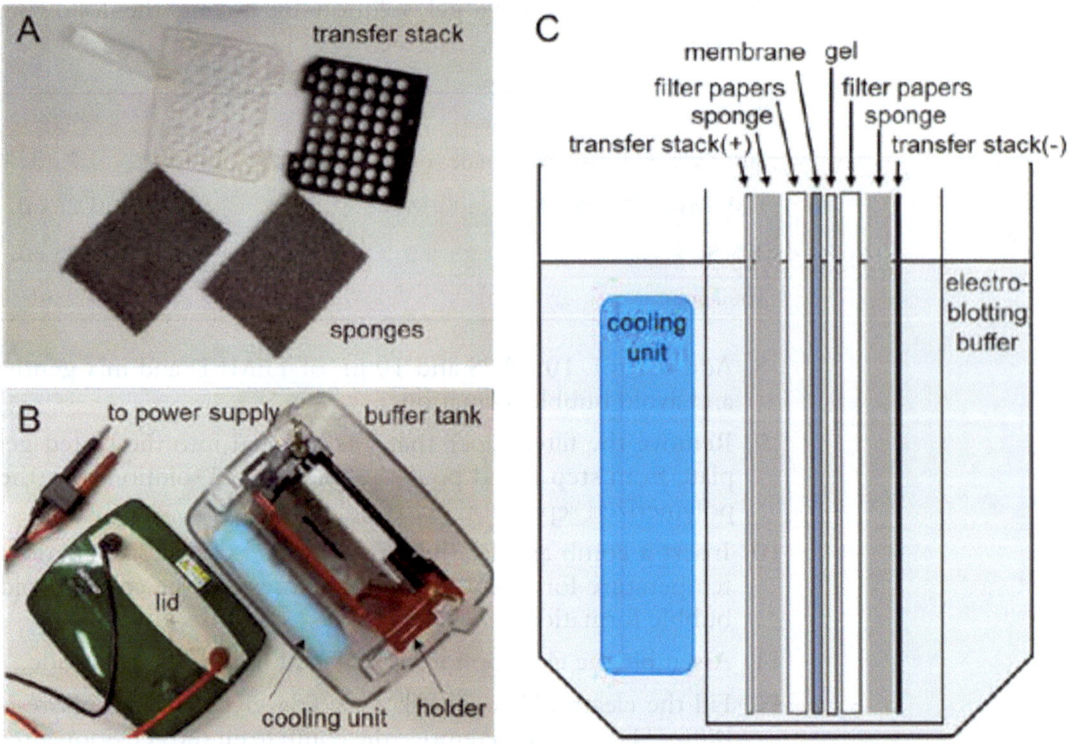

Fig. 1 Protein electroblotting system. (**a**) Transfer stacks and sponges used in a transfer apparatus. (**b**) A transfer apparatus consists of a holder and a tank for grabbing the transfer stack and immersing it in buffer, respectively, a cooler unit to prevent heating of the buffer, and a lid with electrodes. (**c**) Schematic representation of an assembled electroblotting protein system used for transferring proteins from a gel to a membrane (*see* Subheading 3.5, **step 2**)

2. Assemble the electroblotting system according to the manufacturer's instruction (*see* **Note 9**). From the anode to the cathode sides of the transfer stack, place a sponge, followed by three pieces of filter paper, the PVDF membrane, the gel, three pieces of filter paper, and a sponge. Remove any air bubbles formed between filter papers and the PVDF membrane or between the PVDF membrane and the gel. Place this sandwich, wrapped between the transfer stacks, within the holder provided with the transfer apparatus (*see* Fig. 1).
3. Fill the tank with electroblotting buffer and insert the cooling unit.
4. Electroblot proteins from the gel to the membrane at a 200 mA constant current for 2 h.
5. Remove the sandwich from the transfer apparatus.
6. Carefully remove the PVDF membrane with forceps to prevent from tearing.

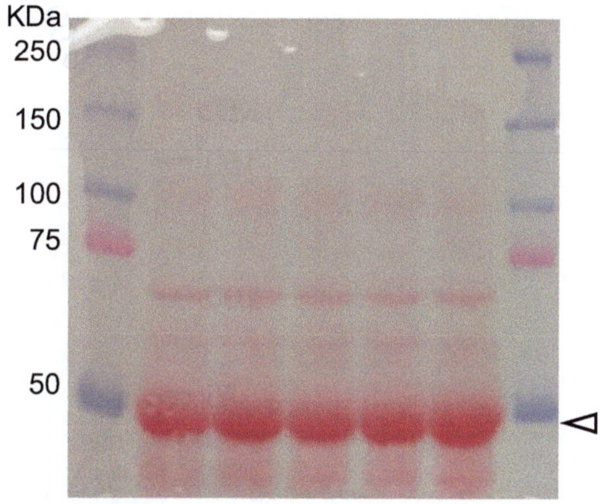

Fig. 2 Blotted membrane stained with Ponceau S. Crude extracts (30 μg) from rice leaf samples collected at different times along a day were separated through SDS–PAGE and then electro-transferred to a PVDF membrane. Strongly stained bands of approximately 50 kD correspond to the Rubisco large subunit (indicated by an arrowhead)

7. Immerse the PVDF membrane in Ponceau S solution and shake by hand for few minutes. Blotted proteins should be visualized as red bands (*see* Fig. 2).
8. Wash the stained membrane with Milli-Q water a few times.
9. Place the stained membrane on the lab bench and cut off the unblotted excess area with a scalpel (*see* **Note 10**).

3.6 Immunodetection

1. Immerse the blotted membrane in blocking solution and gently shake using a seesaw shaker at 4 °C for at least 1 h or preferably overnight (*see* **Note 11**).
2. Discard the solution and immerse the blotted membrane in PBS-T.
3. Gently shake on a seesaw shaker for 10 min at room temperature.
4. Repeat **steps 2** and **3** twice.
5. Discard PBS-T and immerse the blotted membrane in blocking solution containing the primary antibody (1:5000; *see* **Note 12**).
6. Gently shake on a seesaw shaker at 4 °C overnight.
7. Repeat **steps 2** and **3** three times.
8. Discard the solution and immerse the blotted membrane in blocking solution containing the secondary antibody conjugated with HRP (1/10,000 dilution).
9. Gently shake on a seesaw shaker at 4 °C for at least 1 h.
10. Repeat **steps 2** and **3** three times.

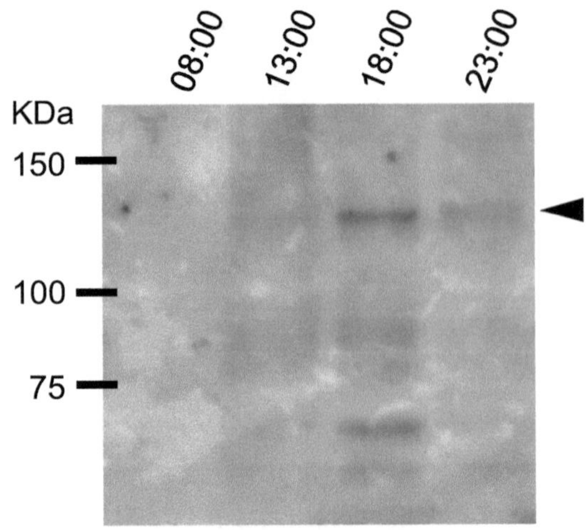

Fig. 3 Immunoblotting for detection of HA:Os–GI in rice plants expressing *HA:Os–GI* (under the regulation of its native promoter). Plants were entrained by short-day conditions (10 h light/14 h dark; subjective dawn at 08:00). Protein blot was probed with an anti-HA antibody. The arrowhead indicates the position of HA:Os–GI and corresponds to the estimated molecular weight of the intact HA:Os–GI protein (around 129.3 kDa). Photoperiod: 08:00–18:00

11. Drain the membrane and eliminate PBS-T residues by gently touching its edges with paper towels. Place the membrane on food wrap on the lab bench or any other flat surface.
12. Prepare the chemiluminescent reagent according to the manufacturer's instructions.
13. Slowly spread the chemiluminescent reagent over the entire membrane.
14. Incubate for the appropriate time according to manufacturer's instructions.
15. Carefully blot away excess reagent using a paper towel and either place the membrane in a hybridization bag or between clear plastic sheets.
16. Visualize chemiluminescent signals using X-ray films or a digital imaging system (*see* Fig. 3 and **Note 13**).

4 Notes

1. *GIGANTEA* (*GI*), a crucial element of the plant circadian clock, is a plant-specific gene whose expression peaks in the evening during daily light–dark cycles. GI has a role in various growth responses, from seedling development to floral transition. HA-tag-fused *Os–GI* gene is expressed under the control of an endogenous 5' *Os–GI* genomic region and can complement the flowering phenotype in an *osgi* mutant [4].

2. This extraction buffer can be stored at 4 °C as stock solution. DNase I and RNase A (1 μL each) as well as one tablet of the protease inhibitor cocktail should be added to 25 mL of extraction buffer immediately before use.
3. The blocking solution can be stored in the refrigerator, at 4 °C, for 1–2 days only.
4. In this experiment, I used leaf blades collected from 2- to 3-week-old plants. Young expanding leaves are suitable for protein extraction because they are soft and easy to grind to a fine powder. When relatively harder materials such as leaves of mature rice plants are used, sea sand should be added during grinding to increase the efficiency of sample destruction. The buffer used for preparing the crude extracts may vary depending on the plant organs used for protein extraction and the nature of the protein of interest. Proteasome inhibitors (e.g., MG-132) or protease inhibitors (e.g., PMSF, phenylmethanesulfonyl fluoride) may also be used. DNase and RNase are optional.
5. Given that the Bradford assay values are modified by the ingredients included in the extraction buffer, the buffer used as blank and for preparing the standard samples should be the same extraction buffer as that used in the preparation of crude extracts. I usually dilute the crude extracts two to five times to obtain absorbance values that fall within the standard curve range. Plotting the absorbance of the samples used as standard results in a linear calibration curve over a certain range. Three replicate samples should be prepared per point, and the average value should be used to calculate the protein concentration.
6. The total protein amounts required for the experiment will depend on the expression levels of the protein under growing conditions and the expression site of the targeted protein in the organ used for extraction as well as to the titer of the antibody used. In addition, regulatory proteins such as transcription factors may have short half-lives. If the extracted protein concentration is low, a large volume of the crude protein extract will be required. In this case, 4× or 5× sample buffers may be used.
7. The quantities shown should be enough for preparing a mini gel (8.5 × 10 cm). For learning about the appropriate concentrations of acrylamide for a better separation of a specific protein, see Ref. [5].
8. The constant current value that should be used may vary with the size of the gel and that of the gel tank you use. When the preceding dyes reach the separation gel, the current may be increased to 30 mA for faster migration of the proteins.

9. I use a Mini Trans-Blot Cell system for electroblotting. In addition to the tank method, a semi-dry method may be used for protein electroblotting. The semi-dry method has the advantage of saving buffer and shortening the time required for blotting. However, the tank method is preferred over the semi-dry method when high molecular protein transfer should be enhanced. For the 30 kDa protein as well as for higher molecular weight proteins, 0.45 μm pore size PVDF membranes are sufficient for blotting. In contrast, for low molecular weight proteins (≤20 kDa), 0.2 μm pore size membrane should be used.
10. If the experiment should be delayed, the membrane may be wrapped in plastic wrap to prevent it from drying out and stored in the refrigerator. I usually move on to the next step the next day.
11. Membrane blocking minimizes nonspecific binding of antibodies to major nonspecific proteins. I usually do this process overnight at 4 °C.
12. Choose primary and secondary antibodies that have been tested in the laboratory. Since nonspecific binding to proteins other than the target protein may occur depending on the concentration of the target protein and the protein composition of the tissues tested, optimal antibody concentrations should be determined in preliminary experiments.
13. If strong or nonspecific immunoblot signals are obtained, antibody concentrations should be decreased. On the contrary, if immunoblot signals are weak, keep the antibody concentration as they were and choose a chemiluminescent reagent with a relatively higher sensitivity. For example, there are three types of Chemi-Lumi One series assay kits: L, Super, and Ultra. As antibodies and chemiluminescent reagents are available from a variety of manufacturers, reagents that are routinely used in each laboratory should be fine.

Acknowledgments

This work was supported by JSPS KAKENHI grant numbers JP19K05983 and JP24K01735.

References

1. Sawa M, Nusinow DA, Kay SA et al (2007) FKF1 and GIGANTEA complex formation is required for length measurement in *Arabidopsis*. Science 318:261–265
2. David KM, Armbruster U, Tama N et al (2006) *Arabidopsis* GIGANTEA protein is post-transcriptionally regulated by light and dark. FEBS Lett 580:1193–1197

3. Simonian MH, Smith JA (2003) Spectrophotometric and colorimetric determination of protein concentration. In: Ausubel FM, Brent R, Kingston RE et al (eds) Current protocols in molecular biology. Wiley, New York
4. Izawa T, Mihara M, Suzuki Y et al (2011) Os-GIGANTEA confers robust diurnal rhythms on the global transcriptome of rice in the field. Plant Cell 23:1741–1755
5. Sambrook J, Fritsch EF, Maniatis T (1989) Detection and analysis of proteins expressed from cloned genes. In: Sambrook J, Fritsch EF, Maniatis T (eds) Molecular cloning, 2nd edn. Cold Spring Harbor laboratory press, New York

Chapter 6

Preparation and Sectioning of Paraffin-Embedded Tissue for Histology and Histochemistry

Masayuki Miya, Ken-ichiro Hibara, and Jun-ichi Itoh

Abstract

Cutting thin sections of plant tissues enables observation of internal structures and inner cell layers, and it facilitates various histochemical staining methods. Paraffin embedding gives the tissues a uniform stiffness that is sufficient for sectioning with a microtome. Here, we describe the procedures for preparation of paraffin-embedded tissues and the techniques for sectioning with a microtome. This method is applicable to various types of tissues, including rice shoot apex and immature inflorescence.

Key words Tissue fixation, Paraffin embedding, Microtome sectioning, Shoot apex, Inflorescence primordia

1 Introduction

Sectioning of tissues is a basic technique required for histological analyses. Biological tissues exhibit heterogeneous stiffness and are difficult to cut into thin sections without prior treatment. There are several methods for sectioning of tissues, including vibratome sectioning of agarose-embedded tissues [1] and microtome sectioning of tissues embedded in paraffin wax [2] or resin [3]. Paraffin embedding and sectioning can be performed on a wide range of plant tissues; it provides sections several micrometers in thickness that are compatible with various types of histochemical analyses, including visualization of cell structure with staining reagents, in situ hybridization for detection of gene expression patterns, and immunolocalization of proteins.

Paraffin embedding begins with dissection and fixation of the tissue of interest. Formalin-aceto-alcohol (FAA) solution and paraformaldehyde (PFA) solution are commonly used as fixatives. FAA solution is superior for tissue infiltration, whereas PFA solution is suitable when preparing samples for in situ hybridization. After fixation, the tissues are dehydrated by passage through a graded

Taiji Kawakatsu (ed.), *Rice: Methods and Protocols*, Methods in Molecular Biology, vol. 2869,
https://doi.org/10.1007/978-1-0716-4204-7_6,

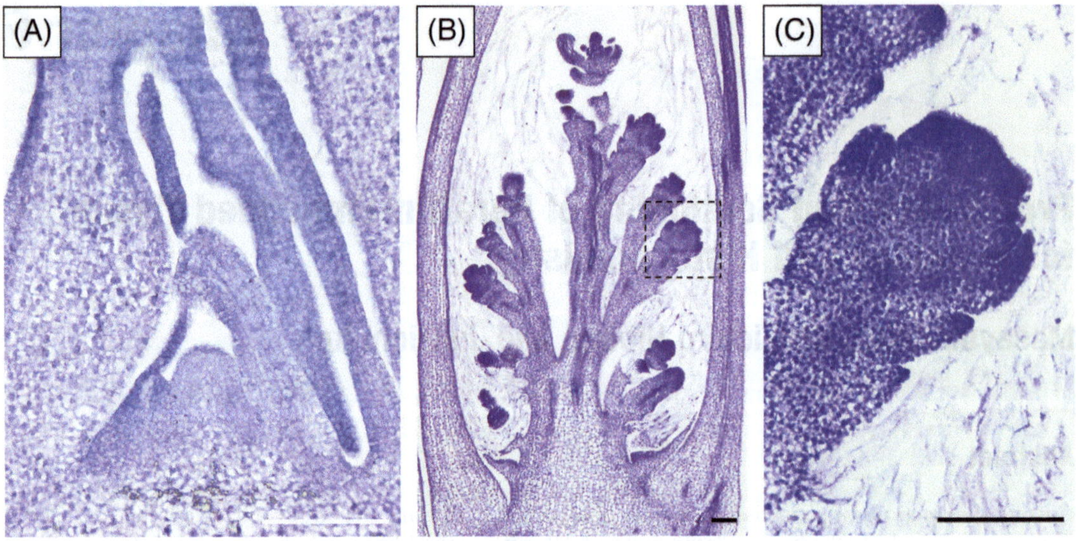

Fig. 1 Sections of paraffin-embedded tissues. (**a**) Shoot apex of 2-week-old rice. (**b**) Immature inflorescence of rice. (**c**) Magnified view of the boxed region in (**b**). Sections were stained with toluidine blue O. Bars: 100 μM

ethanol series. Because paraffin wax is not miscible with ethanol, clearing agents are used as intermediate solvents. Xylene is commonly used as a solvent, but it is toxic and must be handled with care. Another clearing agent, Histo-Clear, is safer and can be directly substituted for xylene. After tissues have been cleared with Histo-Clear, they are infiltrated with paraffin wax for tissue embedding, such as Paraplast Plus. Paraffin-infiltrated tissues can be solidified to prepare paraffin blocks for sectioning with a microtome.

Using this method, we can section rice shoot apex and immature inflorescence. Produced sections can be treated with tissue-staining reagents, such as toluidine blue O (*see* Fig. 1), or used for in situ hybridization analysis if the tissues are fixed with PFA solution (*see* Chap. 7).

2 Materials

2.1 Preparation of Paraffin-Embedded Tissues

1. Formalin-aceto-alcohol (FAA) solution (*see* **Note 1**): Mix 2.5 mL of acetic acid, 5 mL of 37% formaldehyde solution, and 25 mL of 100% ethanol. Make up to 50 mL with distilled water.
2. 1 M phosphate buffer, pH 7.2: Weigh 19.8 g of Na_2HPO_4 and 8.1 g of KH_2PO_4, and then transfer to a 200 mL bottle. Make up to 200 mL with sterilized distilled water. Autoclave and store at room temperature.

3. 4% paraformaldehyde (PFA) solution (*see* **Note 2**): Put 2 g of paraformaldehyde powder in 40 mL of distilled water and add 2 μL of 10 N NaOH. Heat to 60 °C and stir in a fume hood until PFA powder is dissolved (*see* **Note 3**). Add 500 μL of Triton X-100. Cool the solution on ice and add 5 mL of 1 M phosphate buffer, pH 7.2. Make up to 50 mL with sterilized distilled water.
4. 0.1 M phosphate buffer, pH 7.2.
5. Graded ethanol series: 30%, 50%, 70%, 90%, and 100% ethanol. Dilute 100% ethanol with sterile distilled water (*see* **Note 4**).
6. 90% ethanol containing 0.1% (w/v) eosin (*see* **Note 5**).
7. Graded Histo-Clear series: 3:1, 1:1, and 1:3 ethanol/Histo-Clear. Mix 100% ethanol (*see* **Note 4**) and Histo-Clear.
8. Graded Paraplast Plus series: 3:1, 1:1, and 1:3 Histo-Clear/Paraplast Plus. Melt Paraplast Plus in an incubator set at 60 °C (*see* **Note 6**) and mix with Histo-Clear.
9. 5 mL glass vials.
10. Stereomicroscope.
11. Vacuum desiccator.
12. Incubator set at 60 °C.
13. Polystyrene weighing boats.
14. Hot plate.

2.2 Sectioning of Paraffin-Embedded Tissues

1. Rotary microtome.
2. Microtome blades.
3. Microscope slides coated with 3-aminopropyl triethoxysilane.
4. Wooden blocks for microtome sectioning.
5. Steel spatula.
6. Heavy paper.
7. Handsaw.

3 Methods

3.1 Sample Preparation and Fixation

1. Put 4 mL of FAA or PFA solution in 5 mL glass vials (*see* **Note 7**). PFA solution should be kept on ice during sample preparation.
2. Dissect the tissue of interest using tweezers and razor blades under a stereomicroscope, and then put the tissue into the glass vials (*see* **Note 8**).
3. Put the vials in a vacuum desiccator. Switch on the vacuum pump and gradually decrease the pressure until bubbles slowly

emerge from the tissue (*see* **Note 9**). Close the desiccator and switch off the pump. Infiltrate with fixative for 5–10 min.

4. Open the desiccator slightly and slowly to increase the pressure (*see* **Note 10**). Close the vials with the caps and gently invert several times to check whether the tissue sinks in the solution. If the tissue continues to float, repeat **step 3** (*see* **Note 11**).
5. Incubate the vials at 4 °C for 24–48 h (*see* **Note 12**).

3.2 Dehydration

1. If PFA is used as the fixative, exchange PFA solution with 0.1 M phosphate buffer (pH 7.2) and incubate at 4 °C for 1 h.
2. Exchange the solution with a graded ethanol series (increasing concentrations, 30%, 50%, 70%, 90%, and 100%). Incubate the samples at 4 °C for 1 h in each ethanol concentration (*see* **Notes 5** and **13**). This process dehydrates the samples. If FAA is used as the fixative, begin the dehydration process from 50% ethanol.
3. Exchange the 100% ethanol with fresh 100% ethanol and incubate at 4 °C overnight.
4. Exchange the 100% ethanol with fresh 100% ethanol and incubate at room temperature for 1 h.

3.3 Clearing

1. Exchange the ethanol with a graded Histo-Clear series (increasing concentrations, 3:1, 1:1, 1:3). Incubate the samples at room temperature for 1 h in each Histo-Clear concentration.
2. Exchange the solution with 100% Histo-Clear and incubate at room temperature for 1 h.
3. Exchange the Histo-Clear with fresh Histo-Clear and incubate at room temperature overnight.

3.4 Paraffin Infiltration

1. Exchange the Histo-Clear with 3:1 Histo-Clear/Paraplast Plus and incubate the samples at 60 °C for 6 h (*see* **Notes 14** and **15**).
2. Exchange 3:1 Histo-Clear/Paraplast Plus with 1:1 Histo-Clear/Paraplast Plus and incubate the samples at 60 °C for 6 h (*see* **Note 15**).
3. Exchange 1:1 Histo-Clear/Paraplast Plus with 3:1 Histo-Clear/Paraplast Plus and incubate the samples at 60 °C overnight.
4. Exchange 1:1 Histo-Clear/Paraplast Plus with 100% Paraplast Plus and incubate at 60 °C for 6 h (*see* **Note 15**).
5. Repeat **step 4** (*see* **Note 16**).
6. Exchange 100% Paraplast Plus with fresh 100% Paraplast Plus and incubate at 60 °C overnight (*see* **Note 17**)

3.5 Embedding Samples in Paraffin Block

1. Heat the hot plate to 65 °C.
2. Place polystyrene weighing boats on the plate and pour in melted Paraplast Plus to a depth of approximately 1 cm.
3. Place samples in a grid pattern on the bottom of the boats using preheated tweezers. Separate each sample by at least 1 cm.
4. Switch off the hot plate and allow the paraffin wax to solidify. When the paraffin wax begins to solidify and the samples become immobile, float the boats in ice water to allow the paraffin to completely solidify (*see* **Note 18**).
5. Paraffin blocks can be stored at 4 °C for several months (*see* Fig. 2a).

3.6 Sectioning of Paraffin-Embedded Tissues

1. Heat the hot plate to 42 °C.
2. Cut the sample-containing paraffin blocks into pieces with a handsaw, and then arrange the pieces into rectangular prisms using a blade or knife (*see* **Note 19**).

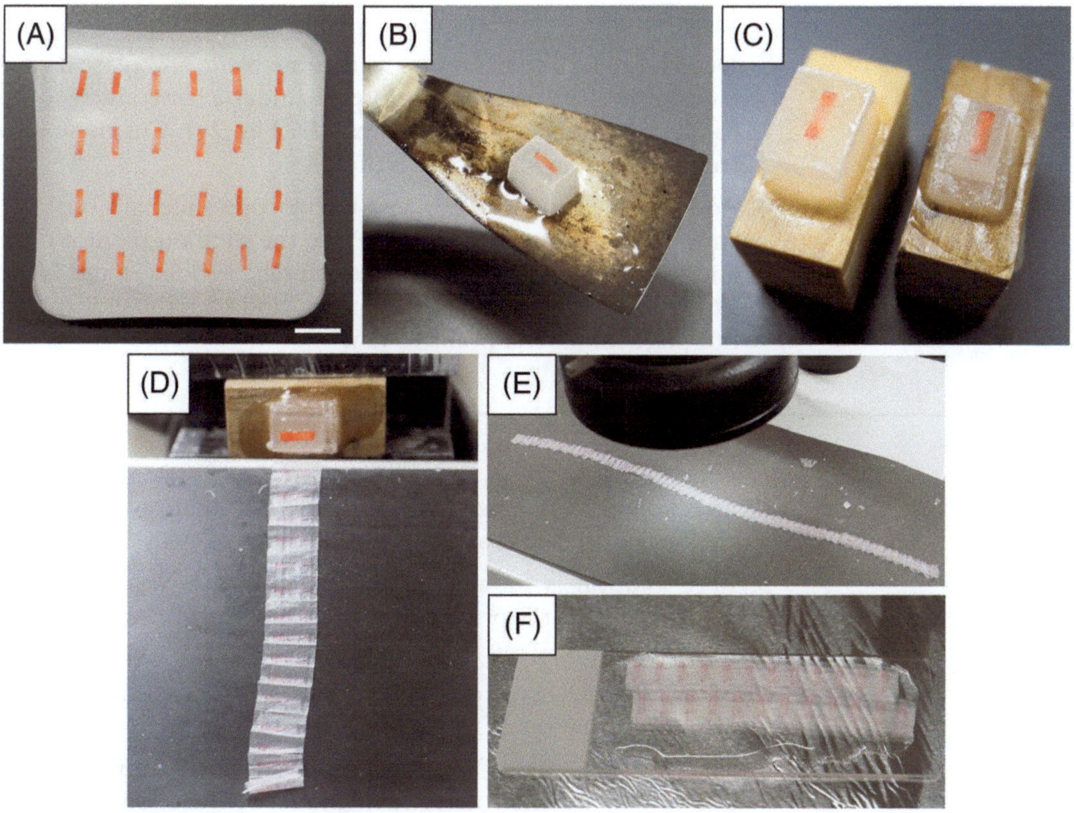

Fig. 2 Sectioning of paraffin-embedded tissue. (**a**) Samples embedded in a paraffin block. (**b**) Melting the surface of the paraffin block using a heated spatula. (**c**) Paraffin blocks adherent on wooden blocks before (left) and after (right) trimming. (**d**) Paraffin ribbon sectioned using a microtome. (**e**) Paraffin ribbon placed on heavy paper and viewed under a stereomicroscope. (**f**) Strips of paraffin ribbon floating on water, on top of a microscope slide placed on the hot plate. Bar in (**a**): 1 cm

3. Heat a steel spatula with a gas flame and bring it into contact with the bottom of the paraffin block to melt the surface (*see* Fig. 2b). Place the paraffin block and melted paraffin wax on top of the wooden block, and then ensure that the paraffin block has firmly adhered to the wooden block.
4. Trim the paraffin block, leaving a margin of 2–3 mm around the sample (*see* Fig. 2c).
5. Fix the wooden block to the microtome with the long side of the paraffin block facing the microtome blade. Adjust the angle of the paraffin block to ensure that the bottom edge of the paraffin block is parallel to the microtome blade. Set the clearance angle of the microtome blade (*see* **Note 20**).
6. Set the section thickness to approximately 60 μm and trim the paraffin block while sectioning until a slice of the sample appears (*see* **Note 21**).
7. Set the section thickness to 8 μM and proceed with sectioning to make a paraffin ribbon (*see* **Note 22**). Pick up the end of the paraffin ribbon using tweezers while sectioning to ensure that the ribbon develops smoothly (*see* Fig. 2d and **Note 23**).
8. Detach the ribbon from the microtome blade and place it on a piece of heavy black paper. Observe under a stereomicroscope to identify the sections of interest (*see* Fig. 2e). Cut these into strips 3–4 cm in length (*see* **Note 24**).
9. Add 1 mL of sterile water to a microscope slide and float the strip of paraffin (*see* **Note 25**). Place the slide on the hot plate and wait 1–2 min for the strip to stretch (*see* Fig. 2f and **Note 26**). Remove the water using a pipette and keep the slide on the hot plate for several hours to dry.

4 Notes

1. FAA solution infiltrates better than PFA solution and works well when preparing samples for histological analysis.
2. PFA solution is suitable when preparing samples for in situ hybridization. It is preferable to prepare fresh PFA solution at the time of use.
3. Avoid excessively increasing the temperature of the solution to prevent formation of formaldehyde gas.
4. The concentration of commercially available ethanol is 99.5%, but it can be used directly without dehydration if passed through a molecular sieve.
5. Staining of samples with eosin improves visibility in the paraffin block. The concentration of eosin should be adjusted depending on the type of tissue.

6. Bubbles may form when Paraplast Plus is melted. Store at 60 °C overnight before use.
7. The volume of fixative should be more than 10× of the tissue volume. Conical tubes can be used, instead of 5 mL vials, when a large amount of tissue must be fixed.
8. It is better to remove the unnecessary parts of the tissue and minimize its size to facilitate infiltration of the fixative.
9. Rapid changes in pressure can cause tissue damage due to sudden enlargement of bubbles in the tissue. Avoid excessively decreasing the pressure to prevent bubbling.
10. Avoid rapid changes in pressure to prevent tissue damage.
11. Sometimes, the tissue floats even after it has been vacuumed several times. In such cases, waiting is recommended because the tissue may sink with penetration of the fixative.
12. The fixation time should be optimized for each type of tissue. If FAA solution is used as the fixative, overnight fixation is sufficient.
13. The tissue can be stored in 70% ethanol for several days. Long-term storage is not recommended because it may cause hardening of the tissue.
14. Do not overheat the tissue because this may cause rapid evaporation of Histo-Clear and insufficient infiltration of paraffin into the tissue. Avoid placing vials directly on metal parts of the incubator because this may transfer excess heat to the tissue.
15. Incubation time may be extended to overnight.
16. Change paraffin rapidly in the incubator to avoid solidification. Complete replacement with new paraffin is not necessary.
17. The tissue may slightly shrink after the entire procedure. Severe shrinkage occurs if paraffin infiltration is insufficient. In such cases, reconsider the tissue sampling method and the infiltration procedure. Reduce the quantity and size of the tissue and extend the time for infiltration.
18. If paraffin is allowed to solidify slowly, bubbles may form around the samples, which may interfere with sectioning by creating holes in the paraffin ribbon. Rapid solidification of paraffin in ice water prevents the formation of such bubbles.
19. To avoid cracking the paraffin block, do not cut excessively large single pieces of paraffin using the blade or knife.
20. In our experience, a clearance angle of 6–8° produces good results, although this depends on the instrument used and the type of tissue.

21. To avoid cracking the paraffin block, do not set the cutting thickness excessively high.
22. The section thickness can be set to 5–15 μM depending on the ease of tissue cutting. Hard tissues can cause the blade to chip, leaving knife marks on the ribbon. If this happens, shift the blade and cut with an unused part of the blade. It is difficult to section tissues composed of especially hard material, such as mature leaves of rice containing silica bodies. In such cases, resin embedding and sectioning would be more suitable.
23. Static electricity may interfere with cutting. Use a static eliminator or humidify the room.
24. The length of the strips should be about two-thirds of the microscope slide length because the strips become longer when stretched.
25. Two or more strips can be mounted if the tissue size allows.
26. To avoid damaging the tissue by overstretching, do not place on the hot plate for an excessive interval. It is sufficient to remove the wrinkles.

Acknowledgments

This work was supported by JSPS KAKENHI grant number 22KJ0772.

References

1. Atkinson JA, Wells DM (2017) An updated protocol for high throughput plant tissue sectioning. Front Plant Sci 8:1721
2. Weigel D, Glazebrook J (2008) Fixation, embedding, and sectioning of plant tissues. CSH Protoc 2008:pdb.prot4941
3. Baum S (2008) Preparation of *Arabidopsis* tissue sections of fixed material. CSH Protoc 2008: pdb.prot4955

Chapter 7

Spatial Analysis of Gene Expression by In Situ Hybridization

Masayuki Miya, Ken-ichiro Hibara, and Jun-ichi Itoh

Abstract

The analysis of the spatial expression patterns of genes is important for deciphering their functions. In situ hybridization provides insight into gene expression patterns at the cellular level. Here we describe a procedure for performing in situ hybridization on sections of paraffin-embedded tissue, including synthesis of labeled RNA probes, hybridization of the probes with target mRNAs, and immunological detection of the signals. The method enables evaluation of the expression patterns of genes in a variety of tissues of rice.

Key words In situ hybridization, Spatial gene expression pattern, RNA probe, Digoxigenin, NBT/BCIP

1 Introduction

The expression of a gene is tightly regulated spatiotemporally and closely linked to its function. Therefore, it is important to analyze the expression patterns of genes to determine their functions. In situ hybridization enables visualization of spatial gene expression patterns in tissues. It can be performed on tissue sections [1, 2] or intact tissue [3, 4], the latter being termed whole-mount in situ hybridization. Because sectioning facilitates the infiltration of reagents into the tissue, in situ hybridization is typically easier to perform on tissue sections than on intact tissues.

The principle of in situ hybridization is the hybridization of labeled probes to target mRNAs and their detection. DNA and RNA probes can be used, but the latter are superior in terms of hybridization stringency. This is because RNA–RNA hybrids are more stable than RNA–DNA hybrids and excess single-stranded RNA probes can be degraded with RNase A, which does not degrade double-stranded RNAs. The various types of probe labels are used in different ways. Radioisotope labeling and autoradiography were formerly performed, but the difficulty handling radioactive material has made their use rare. The most commonly used

Taiji Kawakatsu (ed.), *Rice: Methods and Protocols*, Methods in Molecular Biology, vol. 2869,
https://doi.org/10.1007/978-1-0716-4204-7_7,

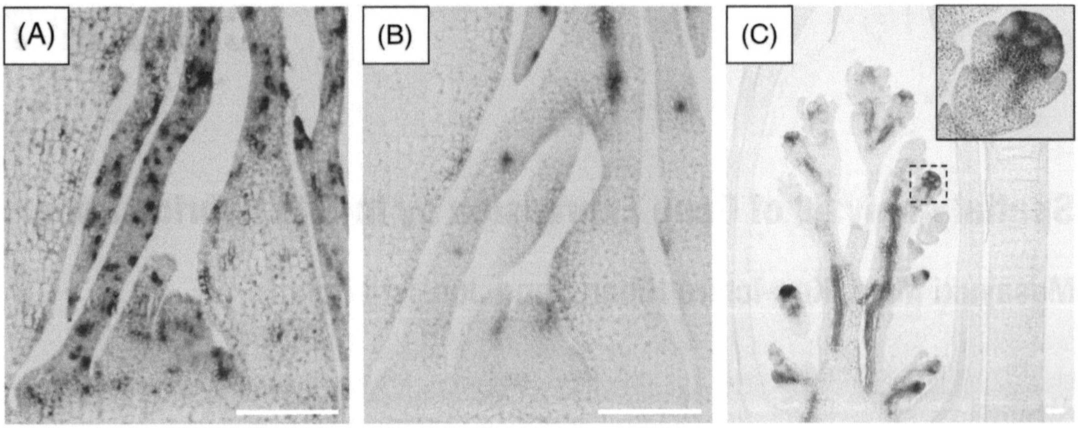

Fig. 1 Signals detected by in situ hybridization using DIG-labeled antisense probes. (**a**) Expression of histone *H4* in shoot apex of rice. (**b**) Expression of *OSHB3* in shoot apex of rice. (**c**) Expression of *OSH1* in immature inflorescence of rice. Inset shows magnified view of the boxed region. Bars, 100 μM

labels are digoxigenin (DIG) and biotin, which are indirectly detected immunologically. Fluorescent-labeled probes are also used and can be detected both fluorescently and immunologically. For immunological detection, alkaline phosphatase (AP)-conjugated antibodies are used to recognize labels and hydrolyze the colorimetric substrate nitroblue tetrazolium (NBT)/5-bromo-4-chloro-3-indolyl-phosphate (BCIP) to an insoluble purple precipitate 3-hydroxy-N-2′-biphenyl-2-naphthalenecarboxamide phosphate ester (HNPP)/Fast Red TR. Immunological and enzyme-labeled detection enables the amplification of weak signals, thereby enhancing sensitivity.

In this chapter, we present an example of in situ hybridization involving DIG-labeled RNA probes and AP-conjugated anti-DIG antibodies to detect signals under visible light. This method can be performed on paraffin-embedded sections of a variety of types of tissue (*see* Chap. 6). The expression patterns of genes are visualized using an optical microscope (*see* Fig. 1).

2 Materials

2.1 In Vitro Transcription of DIG-Labeled RNA Probes

1. DIG RNA labeling kit (SP6 or T7).
2. cDNA of the gene of interest cloned into a vector harboring the SP6 or T7 promotor (*see* **Note 1**).
3. Phenol/chloroform/isoamyl alcohol (25: 24: 1).
4. 0.5 M EDTA, pH 8.0.
5. 4 M lithium chloride.
6. 0.4 M MOPS, pH 7.0.

7. Denaturing gel: Weigh 0.5 g agarose powder and transfer to a 100 mL beaker. Add 44.5 mL distilled water and 2.5 mL 0.4 M MOPS (pH 7.0). Heat and melt the agarose in a microwave and allow it to cool to around 50 °C. Add 3 mL of 37% formaldehyde solution (*see* **Note 2**). Pour the gel into a gel-casting tray and allow it to solidify.
8. RNA loading dye: 50% (v/v) formamide, 6% (w/v) formaldehyde, 0.02 M MOPS (pH 7.0), DNA loading dye (1×), and 81.25 μg/mL ethidium bromide.

2.2 Hybridization of DIG-Labeled RNA Probes

1. Slides prepared by sectioning paraffin-embedded tissues (*see* **Note 3**).
2. Glass staining jars (*see* Fig. 2a): Dry-heat sterilize at 200 °C for 4 h before use.
3. Stainless steel staining rack (*see* Fig. 2a): Dry-heat sterilize at 200 °C for 4 h before use.

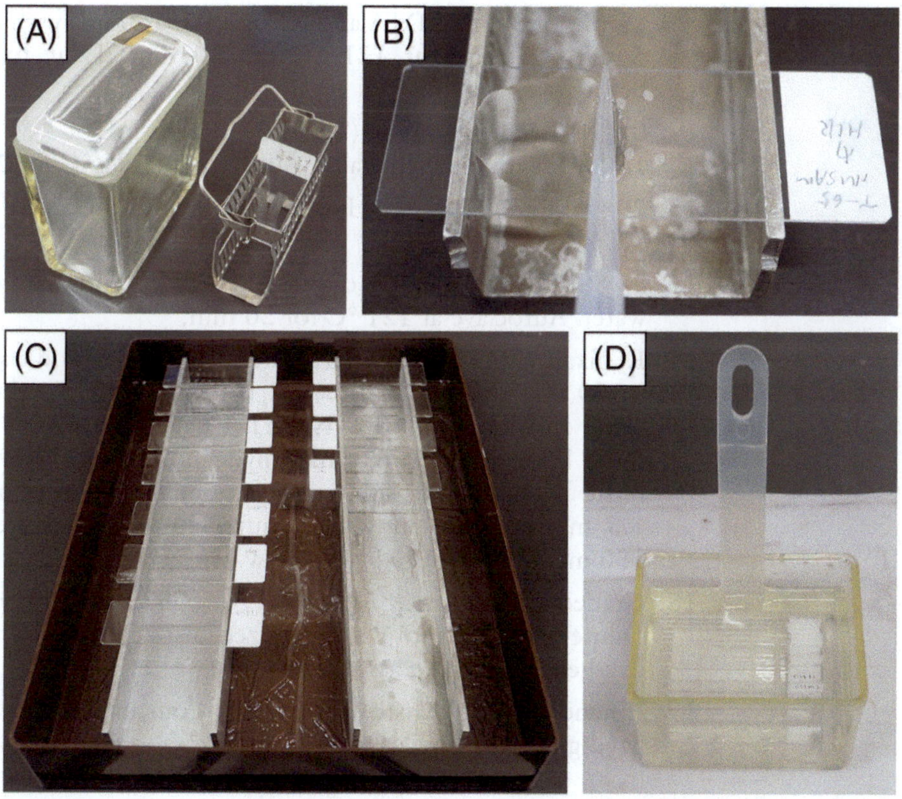

Fig. 2 In situ hybridization procedure. (**a**) A glass staining jar and a stainless steel staining rack. (**b**) Spreading hybridization solution on a microscope slide using the side of a pipette tip. (**c**) Microscope slides placed on aluminum rails in a container. (**d**) Microscope slides on a staining rack with a horizontal tray. The rack is soaked in 1× buffer 1 and gently shaken on a shaker

4. U-shaped aluminum rails (*see* Fig. 2b, c): Dry-heat sterilize at 200 °C for 4 h before use.
5. Tweezers: Dry-heat sterilize at 200 °C for 4 h before use.
6. Container with a lid.
7. Acetic anhydride.
8. Proteinase K (20 mg/mL): Dissolve proteinase K in nuclease-free water and store at −20 °C.
9. tRNA (100 mg/mL): Dissolve tRNA in nuclease-free water and store at −20 °C.
10. Poly(A) (10 mg/mL): Dissolve poly(A) in nuclease-free water and store at −20 °C.
11. TE buffer: 100 mM Tris–HCl, pH 7.5, 10 mM EDTA.
12. Solution A: Mix 25 mL deionized formamide, 3 mL of 5 M NaCl, 5 mL of TE buffer, 1 mL of Denhardt's solution (50×), 10 mL of 50% dextran sulfate solution, 0.5 mL of 3 M DTT, and 0.5 mL of nuclease-free water. Store at −20 °C.
13. Proteinase K buffer: Mix 15 mL of 1 M Tris–HCl (pH 7.5) and 15 mL of 0.5 M EDTA and make up to 150 mL with distilled water. Autoclave at 121 °C for 20 min. Heat to 37 °C before use.
14. 0.1 M Triethanolamine (TEA): Dilute 15 mL of 1 M TEA to 150 mL with distilled water. Autoclave at 121 °C for 20 min.
15. 20× SSPE: 0.2 M NaH_2PO_4, 3 M NaCl, 0.02 M EDTA, pH 7.4.
16. 2× SSPE: Dilute 30 mL of 20× SSPE to 300 mL with distilled water. Autoclave at 121 °C for 20 min.
17. Re-fixation buffer: Weigh 6 g paraformaldehyde and transfer to a 150 mL medium bottle. Add 120 mL of sterilized distilled water and 12 μL of 10 N NaOH. Heat to 60 °C and stir in a fume hood until paraformaldehyde powder is dissolved (*see* **Note 4**). Cool the solution to room temperature and add 1.5 mL of 1 M phosphate buffer (pH 7.2). Make up to 150 mL with sterilized distilled water.
18. Graded ethanol series: 30%, 50%, 70%, 90%, and 100% ethanol. Dilute 100% ethanol with sterilized distilled water. Pour into dry-heat sterilized glass staining jars.
19. Ethanol (100%)/Histo-Clear (1:1): Pour into dry-heat sterilized glass staining jars.

2.3 Immunological Detection of Signals

1. Staining rack with horizontal tray and staining jar (*see* Fig. 2d).
2. Anti-digoxigenin-AP, Fab fragment.
3. NBT/BCIP Stock Solution.

4. Poly Mount.
5. RNase A (10 mg/mL): Dissolve RNase A in nuclease-free water and store at −20 °C.
6. 20× SSC: 3 M NaCl, 0.3 M sodium citrate. Weigh 175 g of NaCl and 88.2 g of trisodium citrate dihydrate. Transfer to a 1 L medium bottle and make up to 1 L with distilled water. Autoclave and store at room temperature.
7. 4× SSC: Dilute 100 mL of 20× SSC to 500 mL with distilled water. Autoclave at 121 °C for 20 min. Heat to 50 °C before use.
8. 0.5× SSC: Dilute 10 mL of 20× SSC to 400 mL with distilled water. Autoclave at 121 °C for 20 min. Heat to 50 °C before use.
9. 10× buffer 1: 1 M Tris–HCl, 1.5 M NaCl, pH 7.5. Weigh 121.14 g of Tris and 87.8 g of NaCl, and transfer to a 1 L beaker. Add 900 mL of distilled water and adjust pH to 7.5 with HCl. Transfer to a 1 L medium bottle and make up to 1 L with distilled water. Autoclave and store at room temperature.
10. 10× buffer 2: 1 M Tris–HCl, 1 M NaCl, pH 9.5. Weigh 121.14 g of Tris and 58.6 g of NaCl, and transfer to a 1 L beaker. Add 900 mL distilled water and adjust pH to 9.5 with HCl. Transfer to a 1 L medium bottle and make up to 1 L with distilled water. Autoclave and store at room temperature.
11. 1 M $MgCl_2$: Weigh 101.5 g of $MgCl_2 \cdot 6H_2O$. Transfer to a 500 mL medium bottle and make up to 500 mL with distilled water. Autoclave and store at room temperature.
12. 1× buffer 1: Dilute 200 mL of 10× buffer 1 to 2 L with distilled water. Autoclave at 121 °C for 20 min.
13. Bovine serum albumin (BSA, 10 mg/mL): Dissolve BSA with buffer 1 (1×) and store at −20 °C.
14. RNase buffer: Mix 100 mL of 5 M NaCl and 10 mL 1 M of Tris–HCl (pH 7.5) and make up to 1 L with distilled water. Autoclave at 121 °C for 20 min. Heat to 37 °C before use.
15. Blocking solution: Weigh 1 g of blocking reagent (Roche) and transfer to a 200 mL beaker. Add 200 mL of buffer 1 (1×) and heat to 50 °C and stir until the blocking reagent is dissolved.
16. 1× buffer 2: Mix 30 mL of 10× buffer 2 and 15 mL of 1 M $MgCl_2$. Make up to 300 mL with distilled water.

3 Methods

Carry out all procedures at room temperature unless otherwise specified.

3.1 In Vitro Transcription of DIG-Labeled RNA Probes

1. Linearize the plasmid containing cDNA using a restriction enzyme (*see* **Note 5**).
2. Purify the linearized plasmid by phenol/chloroform extraction and ethanol precipitation. Add 200 μL of phenol/chloroform to 200 μL of linearized plasmid solution and mix by vortex mixing. Centrifuge the mixture for 2 min at 20,000× *g* at 4 °C and transfer the aqueous layer to a new tube. Add 200 μL of chloroform and repeat centrifugation and transfer. Add 30 μL of 3 M sodium acetate solution and mix by pipetting. Add 650 μL of 100% ethanol, mix by inverting, and incubate at −80 °C for 10 min. Centrifuge the mixture for 10 min at 20,000× *g* at 4 °C and remove the supernatant. Add 500 μL of 70% ethanol and incubate at room temperature for 10 min. Centrifuge the mixture for 5 min at 20,000× *g* at 4 °C and remove the supernatant. Dry the pellet and dissolve it in 20 μL of nuclease-free water.
3. Mix 1 μg of the linearized plasmid (*see* **Note 6**), 2 μL of 10× NTP labeling mixture, 2 μL of 10× transcription buffer, 1 μL of RNase inhibitor, and 2 μL of T7 or SP6 RNA polymerase on ice and make up to 20 μL with nuclease-free water.
4. Incubate at 37 °C for 2 h.
5. Stop the reaction by adding 1 μL of 0.5 M EDTA (*see* **Note 7**).
6. Purify the synthesized RNA probe by lithium chloride precipitation. Add 2.5 μL of 4 M lithium chloride and mix by pipetting. Add 65 μL of 100% ethanol, mix by inverting, and incubate at −80 °C for 30 min. Centrifuge the mixture for 10 min at 20,000× *g* at 4 °C and remove the supernatant. Add 500 μL of 70% ethanol and incubate at room temperature for 10 min. Centrifuge the mixture for 5 min at 20,000× *g* at 4 °C and remove the supernatant. Dry the pellet and dissolve it in 100 μL of nuclease-free water containing 1% (v/v) RNase inhibitor (*see* **Note 8**).
7. Mix 2.5 μL of RNA probe solution with 8 μL of RNA loading dye.
8. Incubate in a heating block set at 60 °C for 5 min, and cool on ice for 5 min.
9. Load 10.5 μL of RNA probe solution with RNA loading dye in a denaturing gel and electrophorese in 0.4 M MOPS buffer (pH 7.0). Confirm that a clear band appears (*see* **Note 9**). Store the RNA probe at −80 °C.

3.2 Hybridization of DIG-Labeled RNA Probes

1. Place microscope slides in a staining rack (*see* Fig. 2a).
2. Transfer the staining rack to a glass staining jar with Histo-Clear, and incubate for 10 min.
3. Incubate in fresh Histo-Clear for 10 min.

4. Incubate in 1:1100% ethanol/Histo-Clear for 5 min.
5. Incubate in 100% ethanol for 5 min.
6. Repeat **step 5** with fresh 100% ethanol.
7. Incubate in fresh 100% ethanol for 2 min.
8. Incubate in 90% ethanol for 2 min.
9. Incubate in 70% ethanol for 2 min.
10. Incubate in 50% ethanol for 2 min.
11. Incubate in 30% ethanol for 2 min.
12. Incubate in sterilized distilled water for 5 min.
13. Repeat **step 12** with fresh sterilized distilled water.
14. Add 37.5 μL of 20 mg/mL proteinase K to 150 mL of proteinase K buffer prewarmed to 37 °C. Incubate the slides in the resulting proteinase K solution at 37 °C for 20 min (*see* **Note 10**).
15. Repeat **step 12** twice with fresh sterilized distilled water.
16. Incubate in re-fixation buffer for 10 min.
17. Repeat **step 12** twice with fresh sterilized distilled water.
18. Transfer the staining rack to an empty glass staining jar. Add 750 μL of acetic anhydride to 150 mL of 0.1 M TEA and mix for several seconds. Pour the solution into the jar and incubate for 10 min (*see* **Note 11**).
19. Incubate in 2× SSPE for 5 min.
20. Repeat **step 19** with fresh 2× SSPE.
21. Incubate in 30% ethanol for 2 min (*see* **Note 12**).
22. Incubate in 50% ethanol for 2 min (*see* **Note 12**).
23. Incubate in 70% ethanol for 2 min (*see* **Note 12**).
24. Incubate in 90% ethanol for 2 min (*see* **Note 12**).
25. Incubate in 100% ethanol for 5 min (*see* **Note 12**).
26. Repeat **step 25** with fresh 100% ethanol.
27. Place the staining rack in the vacuum desiccator. Vacuum dry for >1 h.
28. Make 30 μL of probe solution per slide by mixing 3 μL of 3 M DTT, 3 μL of 100 mg/mL tRNA, 15 μL of 10 mg/mL poly (A), 8 μL of nuclease-free water, and 1 μL of DIG-labeled RNA probe (*see* **Note 13**). Incubate in a heating block set at 80 °C for 5 min, and cool on ice for >2 min.
29. Make 300 μL of hybridization solution per slide by mixing 30 μL of probe solution with 270 μL of solution A (*see* **Note 14**). Store in an incubator set at 50 °C.

30. Place paper wipes in the container and wet them with 40% formamide. Place aluminum rails on the paper wipes.
31. Place 300 μL of hybridization solution on the slide and spread the solution using the side of the pipette chip so that it covers all of the tissue sections (*see* Fig. 2b and **Note 15**). Remove bubbles using the pipette.
32. Place a cover glass on the slide and lay it on the aluminum rails (*see* Fig. 2c). Put the lid on the container.
33. Place the container in an incubator set at 50 °C and incubate for >16 h (*see* **Note 16**).

3.3 Immunological Detection of Signals

1. Pour 4× SSC into a glass staining jar. Immerse the slide in 4× SSC and slip down the cover glass in 4× SSC (*see* **Note 17**). Place the slides in the staining rack using a horizontal tray.
2. Incubate in 4× SSC for 15 min at 50 °C.
3. Incubate in RNase buffer for 5 min at 37 °C.
4. Add 1 mL of 10 mg/mL RNase A to 200 mL of RNase buffer. Incubate the slides in RNase A solution for 30 min at 37 °C.
5. Incubate in RNase buffer for 5 min at 37 °C.
6. Repeat **step 5** twice with fresh RNase buffer.
7. Incubate in 0.5× SSC for 20 min at 50 °C.
8. Repeat **step** 7 with fresh 0.5× SSC.
9. Incubate in 1× buffer 1 for 5 min.
10. Repeat **step 9** with fresh 1× buffer 1.
11. Incubate in blocking solution for 30 min.
12. Place paper wipes in the container and wet them with 1× buffer 1. Place aluminum rails on the paper wipes.
13. Make 500 μL of antibody solution per slide by mixing 450 μL of 1× buffer 1, 50 μL of 10 mg/mL BSA, and 0.5 μL of anti-digoxigenin-AP. Wipe the back side of the slide and place it on an aluminum rail. Place 500 μL of antibody solution on the slide so that it covers all of the tissue sections. Put the lid on the container and incubate for 60 min.
14. Remove the antibody solution. Transfer the slides to the staining rack using a horizontal tray.
15. Incubate in 1× buffer 1 for 10 min with gentle shaking (*see* Fig. 2d).
16. Repeat **step 15** twice with fresh 1× buffer 1.
17. Incubate in 1× buffer 3 for 5 min.
18. Make 1 mL of NBT/BCIP solution per slide by mixing 992 μL of 1× buffer 2 and 8 μL of NBT/BCIP stock solution. Wipe the back side of the slide and place it on an aluminum rail. Place

1 mL of NBT/BCIP solution on the slide so that it covers all of the tissue sections. Put the lid on the container and incubate at 37 °C for 3 h (*see* **Note 18**).

19. Remove NBT/BCIP solution and check for staining under a microscope (*see* **Note 19**). Place the slides in the staining rack.
20. Incubate in distilled water for 2 min.
21. Repeat **step 20** with fresh distilled water.
22. Incubate in 30% ethanol for 1 min (*see* **Note 20**).
23. Incubate in 50% ethanol for 1 min.
24. Incubate in 70% ethanol for 1 min.
25. Incubate in 90% ethanol for 1 min.
26. Incubate in 100% ethanol for 2 min.
27. Repeat **step 26** with fresh 100% ethanol.
28. Incubate in 1:1100% ethanol/Histo-Clear for 2 min.
29. Incubate in Histo-Clear for 2 min.
30. Repeat **step 29** with fresh Histo-Clear.
31. Remove the slides from the staining rack.
32. Cover the slides with Poly Mount and gently apply a coverslip.

4 Notes

1. We typically use 500 to 1000 bp cDNA fragments as templates for probes. A cDNA fragment containing the 3′-UTR region is preferred because of a high sequence specificity for each gene. If a vector without the SP6 or T7 promotor is used, amplify the PCR product using primers containing the promotor and use it directly as the template instead of the linearized plasmid.
2. Do not add formaldehyde solution until it has cooled to prevent formation of formaldehyde gas.
3. Sections should be mounted on a microscope slide coated with an antistripping agent to prevent detachment of the tissue during the procedure.
4. Do not heat the solution excessively to prevent formation of formaldehyde gas.
5. For antisense probes, cut upstream of the 5′-end of the cDNA and transcribe using the promotor downstream of the 3′-end. For sense probes, cut downstream of the 3′-end of the cDNA and transcribe using the promotor upstream of the 5′-end. Use a restriction enzyme that creates 5′-overhangs or blunt ends. Check for complete digestion by electrophoresis.

6. If you use the PCR product instead of the linearized plasmid, use 200 ng of purified PCR product.
7. As an optional step, add 2 μL of RNase-free DNase I and incubate at 37 °C for 15 min to remove template DNA before adding 0.5 M EDTA.
8. Check the concentration of the RNA probe. Approximately 100 ng/μL or more is preferable, but a lower concentration is acceptable if a band is visible after electrophoresis.
9. Sometimes two or more bands appear, which can be caused by remaining DNA template or incomplete digestion of the plasmid. So long as the band of the RNA probe is bright, the number of bands is unimportant.
10. Proteinase K treatment increases the accessibility of the probes to the mRNA, although it can weaken the tissue. The duration of proteinase K treatment should be adjusted for the type of tissue.
11. This acetylation step reduces electrostatic nonspecific binding of the probe to the tissue or slide. Mix and pour the solution quickly because the acetylation ability weakens rapidly.
12. The ethanol series used in **steps 5–11** can be reused.
13. If the concentration of the RNA probe is <100 ng/μL, increase the volume of the RNA probe solution and decrease that of nuclease-free water.
14. Mix the solution thoroughly by vortex mixing.
15. Take care not to scrape the tissue. Slowly move the chip slightly above the tissue to spread the solution by surface tension.
16. The hybridization temperature should be adjusted for each probe and can be 4–60 °C. A higher temperature can yield good results for probes with high GC contents by increasing the hybridization stringency.
17. The cover glass typically slips down easily. If the glass adheres tightly to the slide, do not pull it off because the tissue may be damaged. Immerse the slide in 4× SSC and wait until the solution penetrates between the cover glass and the slide, and then separate them.
18. It is possible to lower the temperature, which slows the staining reaction.
19. If the staining is weak, return the staining solution and incubate for several hours or overnight.
20. Incubation in the ethanol series reduces background staining, although it also weakens signals. If signals are faint, the incubation period can be shortened. It is also possible to skip all subsequent steps and mount the slides in 50% glycerol.

Acknowledgments

This work was supported by JSPS KAKENHI grant number 22KJ0772.

References

1. Brewer PB, Heisler MG, Hejátko J et al (2006) *In situ* hybridization for mRNA detection in *Arabidopsis* tissue sections. Nat Protoc 1:1462–1467
2. Zöllner NR, Bezrutczyk M, Laureyns R et al (2021) An RNA *in situ* hybridization protocol optimized for monocot tissue. STAR Protoc 2: 100398
3. Hejátko J, Blilou I, Brewer PB et al (2006) *In situ* hybridization technique for mRNA detection in whole mount *Arabidopsis* samples. Nat Protoc 1:1939–1946
4. Rozier F, Mirabet V, Vernoux T et al (2014) Analysis of 3D gene expression patterns in plants using whole-mount RNA *in situ* hybridization. Nat Protoc 9:2464–2475

Chapter 8

Purification of Anti-OSH1 IgG and Immunohistochemical Staining of Rice Shoot Apical Meristem

Takumi Tezuka, Misuzu Nosaka-Takahashi, Takanori Yoshikawa, and Yutaka Sato

Abstract

Immunohistochemical staining (IHC) enables the visualization of protein localization within tissues and organs. Here, we describe IHC to detect the ORYZA SATIVA HOMEOBOX1 (OSH1) protein in sections of rice shoot apical meristems (SAMs). First, anti-OSH1 IgG is purified from OSH1 polyclonal rabbit antisera. For subsequent IHC, rice shoot apices are fixed and embedded in paraffin and sections are prepared for IHC. In this IHC procedure, two different labeling methods, fluorescein isothiocyanate and alkaline phosphatase, are utilized for secondary antibody detection.

Key words IHC, IgG purification, Paraffin section, FITC, Alkaline phosphatase

1 Introduction

Immunohistochemical staining (IHC), a method developed by pathologist Coons et al. in 1941 [1], enables the detection of antigen localization, such as proteins, in tissue sections through specific antigen–antibody reactions. Since its inception, numerous advancements have enhanced the IHC technique, rendering it indispensable in pathology for clinically diagnosing infectious and tumorous diseases [2].

IHC can be performed either by the direct method, which uses a primary antibody directly conjugated to a label, or the indirect method, which uses a primary antibody that recognizes the antigen and a labeled secondary antibody that recognizes the primary antibody. The indirect method has superior detection sensitivity compared to the direct method. This is attributed to that the unlabeled primary antibodies generally possess higher activity of antigen recognition than those directly labeled. Moreover, multiple molecules of secondary antibody bind to a single molecule of primary antibody to amplify the signal [3].

Taiji Kawakatsu (ed.), *Rice: Methods and Protocols*, Methods in Molecular Biology, vol. 2869,
https://doi.org/10.1007/978-1-0716-4204-7_8,

IHC is typically conducted by reacting antibodies on tissue sections. However, an alternative approach, whole-mount IHC, employs the fixed entire sample. Whole-mount IHC has the advantage of preserving the three-dimensional structure of tissues, organs, and organisms, and it is commonly employed in animal studies [4]. When preparing sections for IHC, either frozen sectioning or paraffin sectioning is employed [5, 6]. Frozen sections excel in preserving epitope structure and enzyme activity compared to paraffin sections. On the other hand, paraffin sections offer the advantage of better preserving tissue structure compared to frozen sections [6]. Two staining methods, utilizing fluorescein isothiocyanate (FITC) and alkaline phosphatase (AP), are commonly employed to detect secondary antibodies. In the method using secondary antibodies labeled with FITC, the fluorescence of FITC is observed using confocal laser scanning microscopy or fluorescence microscopy. In the method using secondary antibodies labeled with AP, visible blue/purple pigment (NBT formazan) developed in the chemical reaction catalyzed by AP is observed using optical microscopy.

To observe the localization of the OSH1 protein in rice shoot apical meristems (SAMs), we describe herein the process of anti-OSH1 IgG purification from sera, plant sample fixation, and paraffin sectioning, followed by the detection of OSH1 localization using the indirect IHC. Since the initial anti-OSH1 rabbit antiserum contains a variety of antibodies, we purify anti-OSH1 IgG using His-tagged OSH1 protein captured by TALON resin to increase the specificity of IHC signals.

OSH1 is the rice ortholog gene of maize *KNOTTED1* (*KN1*). Both *KN1* and *OSH1* exhibit expression within shoot meristems but undergo downregulation at the flank of the shoot meristem where leaf primordia differentiate [7–9]. Furthermore, in the loss-of-function *osh1* mutant, SAM development is arrested after germination [10]. This observation suggests the indispensable role of *OSH1* in the maintenance of undifferentiated cells within the shoot meristem during vegetative growth. Moreover, *OSH1* expression is repressed in the region where differentiation of leaf primordia occurs. Hence, observing the localization of OSH1 enables the capture of dynamics of shoot meristem activity. OSH1 is detected by IHC exclusively in undifferentiated cells, while nuclei staining with propidium iodide (PI) marks all cells. This differential staining pattern allows for the distinction between undifferentiated and differentiated cells in the SAM. Observation of the number and arrangement of these cells, following the stages of each leaf primordium, revealed a process of replenishment of undifferentiated cells consumed by leaf primordium formation. This elucidation contributes to our comprehension of shoot meristem homeostasis [11].

2 Materials

Sterile distilled water should be used when preparing solutions.

2.1 Antibody Purification from Antiserum

1. *E. coli* BL21 strain transformed with the pET-32-OSH1 vector. The pET-32-OSH1 harbors the complete coding sequence of OSH1, which is in frame with thioredoxin, a 6xHis-tag, and an S-tag for purification purposes.
2. Terrific Broth (TB) liquid medium: Autoclave at 121 °C for 20 min and store at room temperature. Add ampicillin to a final concentration of 50 μg/mL before use.
3. Shaker.
4. Spectrophotometer.
5. 1 M isopropyl β-D-thiogalactopyranoside (IPTG).
6. Large volume centrifuge.
7. BugBuster Protein Extraction Reagent (Merck).
8. Benzonase Nuclease (Merck)
9. Lysis buffer: 20 mM Tris–HCl, pH 8.0, 100 mM NaCl, 8 M urea.
10. Elute buffer: 20 mM Tris–HCl, pH 8.0, 100 mM NaCl, 8 M urea, 100 mM imidazole.
11. TALON affinity resin (Takara).
12. Falcon tube.
13. Rotator.
14. Empty column: TALON 2 mL Disposable Gravity Columns (Takara).
15. Power supply.
16. Protein electrophoresis system: Mini-PROTEAN Tetra Cell (Bio-Rad) or equivalent.
17. Tris–glycine electrophoresis buffer (5x): 125 mM Tris-HCl, 960 mM glycine, 0.5% SDS. Dilute to 1x with sterile water before use.
18. SDS-PAGE gel (*see* **Note 1**).
19. Protein electroblotting system: Mini Trans-Blot Cell (Bio-Rad) or equivalent.
20. Electroblotting buffer: 25 mM Tris-HCl, 192 mM glycine, 20% (v/v) methanol.
21. Filter paper: Whatman cellulose chromatography paper 3 MM Chr.
22. Immobilon-P PVDF membrane, 0.45 μM pore size (Millipore).

23. Anti-OSH1 rabbit antiserum [11].
24. TBS-T: 20 mM Tris–HCl, pH 7.4, 150 mM NaCl, 0.05% (v/v) Tween 20.
25. Antibody elution buffer: 0.1 M glycine-HCl, pH 3.0, 0.15 M NaCl.
26. Neutralizing solution: 0.5 M HEPES-KOH, pH 8.5.
27. 10% sodium azide solution.

2.2 Fixation, Paraffin Embedding, and Sectioning of Plant Sample

1. Fixative solution: 4% (w/v) formaldehyde, 5% (v/v) acetic acid, 50% (v/v) ethanol. Prepare just before use and store at 4 °C.
2. Desiccator.
3. Aspirator.
4. Graded ethanol series: 70%, 85%, and 95% ethanol. Dilute 100% ethanol with sterile distilled water. Store at 4 °C.
5. 100% ethanol: Dehydrate 99.5% ethanol by adding the dried 3A molecular sieves.
6. Histo-Clear II.
7. 1:2 Histo-Clear II/ethanol.
8. 1:1 Histo-Clear II/ethanol.
9. Paraffin wax: Melt Paraplast Plus by incubating at 60 °C.
10. Hot plate.
11. Rotary microtome.
12. MAS Adhesive slide (*Matsunami Glass*).

2.3 IHC

1. Staining rack.
2. Staining jar.
3. Histo-Clear II.
4. 1:1 Histo-Clear II/ethanol.
5. Graded ethanol series: 30%, 50%, 70%, 85%, 90%, and 95% ethanol for FITC staining or 30%, 60%, 80%, and 95% ethanol for AP staining. Dilute 100% ethanol with sterile distilled water.
6. PBS: 0.13 M NaCl, 7 mM Na2HPO4, 3 mM NaH2PO4•H2O.
7. Proteinase K solution: 0.1 mg/mL proteinase K in PBS.
8. Humidity chamber: Light-blocking box that can maintain high humidity. Place a paper towel moistened with sterile distilled water on the bottom and place a rail on top to prevent the slide from touching the wet towel directly.
9. Coverslip.
10. (For FITC staining) Blocking solution: 1% bovine serum albumin, 0.05% Tween 20 in PBS.

11. (For FITC staining) Secondary antibody: FITC Goat Anti-Rabbit IgG (H + L) Conjugate (ZyMax Grade) (Invitrogen).
12. (For FITC staining) PBS-T: PBS with 0.05% Tween 20.
13. (For FITC staining) VECTASHIELD (Vector Laboratories).
14. (For FITC staining) Confocal laser scanning microscopy.
15. (For AP staining) Blocking solution: 0.01% bovine serum albumin, 0.05% Tween 20 in PBS.
16. (For AP staining) Secondary antibody: Secondary Antibody Solution Alkali-Phosphatase Conjugated (Anti-Rabbit) (Invitrogen).
17. (For AP staining) TNM: 100 mM Tris–HCl, pH 9.5, 100 mM NaCl, 50 mM $MgCl_2$.
18. (For AP staining) AP substrate: Add 90 μL of NBT (Roche) and 70 μL of BCIP (Roche) to 20 mL of TNM.
19. (For AP staining) TE: 10 mM Tris–HCl (pH 8.0), 1 mM EDTA.
20. (For AP staining) Lemosol (Fujifilm).
21. (For AP staining) 1:1 Lemosol/ethanol.
22. (For AP staining) EUKITT mounting medium (ORSAtec).
23. (For AP staining) Optical microscope.

3 Methods

Unless otherwise specified, all experiments are carried out at room temperature.

3.1 Antibody Purification from Antiserum

1. Inoculate a single colony of *E. coli* BL21 strain, which has been transformed with the pET-32-OSH1 vector, into 2 mL of TB liquid medium.
2. Incubate at 37 °C with a horizontal shaking at 220 rpm in a shaker until the medium exhibits turbidity.
3. Transfer the entire 2 mL of culture to 10 mL of fresh TB liquid medium.
4. Incubate at 37 °C with a horizontal shaking at 150 rpm in a shaker until the medium exhibits turbidity.
5. Transfer the entire 10 mL of culture to 1 L of fresh TB liquid medium.
6. Incubate at 37 °C with a horizontal shaking at 150 rpm in a shaker. Monitor the absorbance at 600 nm (OD600) using a spectrophotometer. Grow the culture until OD600 reaches 0.600.

7. Add 1 mL of 1 M IPTG.
8. Incubate at 37 °C with a horizontal shaking at 150 rpm in a shaker for 2 h in a shaker.
9. Dispense 250 mL of the culture into four centrifuge tubes.
10. Centrifuge at 2500× *g* at 4 °C for 10 min.
11. Discard the supernatant.
12. Add 10 mL of BugBuster Protein Extraction Reagent to each tube and suspend the sediment by pipetting.
13. Add 10 μL of Benzonase nuclease to each tube and mix by inverting.
14. Incubate at room temperature for 20 min for cell lysis and genomic DNA removal.
15. Transfer the solution to two 50 mL Falcon tubes and centrifuge at 10,000× *g* at 4 °C for 20 min.
16. Discard the supernatant (*see* **Note 2**).
17. Add 20 mL of lysis buffer to the pellet and suspend by pipetting.
18. Centrifuge at 10,000× *g* at room temperature for 10 min.
19. Collect the supernatant and keep on ice.
20. Dispense 2 mL of TALON affinity resin into one Falcon tube (50 mL). Centrifuge at 400× *g* at room temperature for 2 min. Carefully remove the supernatant.
21. Add 10 mL of lysis buffer to Falcon tube from **step 20** for equilibration. Thoroughly mix by pipetting. Centrifuge at 400× *g* at room temperature for 2 min. Carefully remove the supernatant.
22. Add all the solution from **step 19** to equilibrated TALON affinity resin from **step 21**.
23. Rotate by rotator at room temperature for 20 min to capture His-tagged OSH1 protein by TALON affinity resin.
24. Centrifuge at 400× *g* at room temperature for 5 min.
25. Remove the supernatant.
26. Add 20 mL of lysis buffer.
27. Rotate by rotator at room temperature for 10 min.
28. Centrifuge at 400× *g* at room temperature for 5 min.
29. Remove the supernatant.
30. Repeat **steps 26–29**.
31. Add 1 mL of lysis buffer and resuspend. Store the tube on ice.
32. Set the empty columns and wash with 3 mL of lysis buffer.

33. Add the OSH1 protein-bound TALON affinity resin from **step 31** to the column and allow the lysis buffer to flow out.
34. Add lysis buffer to fill the column and allow the lysis buffer to flow out.
35. Add a total of 4.5 mL of elute buffer to the column in several portions and collect the flowthrough. Proceed to the next step or store at −80 °C.
36. Perform SDS-PAGE to separate proteins and transfer to PVDF membrane (*see* Chap. 5 and **Note 3**).
37. Put PVDF membrane in a plastic bag. Add OSH1 polyclonal antisera without dilution in the bag and shake overnight in a shaker at room temperature.
38. Place the PVDF membrane in a square dish and wash it with enough TBS-T to completely immerse the membrane, gently shaking the dish at room temperature for 10 min.
39. Repeat **step 38** five times.
40. Immerse the PVDF membrane in 2 mL of antibody elution buffer in a square dish for 1 min.
41. Collect the antibody eluate and add 450 μL of neutralizing solution.
42. Repeat **steps 39** and **40** five times. Keep the antibody eluates separately at this point.
43. Measure the absorbance of the antibody eluates at 280 nm by spectrophotometer and identify the antibody elutes with the first and second highest absorbance. Add 10% sodium azide as a preservative to achieve a final concentration of 0.05% and store at 4 °C.

3.2 Fixation, Paraffin Embedding, and Sectioning of Plant Sample

1. Collect the basal part of the shoots containing SAM from rice seedlings (*see* Fig. 1) and immerse in a vial containing 10 mL of fixative solution on ice.
2. Place the vial in a desiccator and vacuum it for 30 min using an aspirator (*see* **Note 4**).
3. Incubate the vial at 4 °C overnight.
4. Substitute the fixative solution with a series of ethanol solutions of increasing concentration (70%, 85%, 95%, and 100%) for sample dehydration. For each ethanol concentration, gently shake the vial for 1 h on ice.
5. Gently shake the vial at room temperature overnight.
6. Substitute the 100% ethanol with fresh 100% ethanol, 1: 2 Histo-Clear II/ethanol, 1:1 Histo-Clear II/ethanol, and 100% Histo-Clear II in this order. For each solution, gently shake the vial at room temperature for 1 h.

Fig. 1 Collecting the basal part of the shoot containing the SAM from a 10-day--old seedling. (**a**) A 10-day-old rice seedling with a white box indicating the sample area, approximately 0.5–1 cm from the basal part of the seedling. (**b**) The basal part of the seedling is cut off for sampling, followed by immersion in the fixative solution. Bars indicate 1 cm

7. Substitute the 100% Histo-Clear II with fresh 100% Histo-Clear II twice. For each solution, gently shake the vial at room temperature for 1 h.
8. Remove the 100% Histo-Clear II and fill the vial halfway with fresh 100% Histo-Clear II. Fill the remaining half of the vial with solid paraffin wax.
9. Incubate the vial at room temperature overnight.
10. Incubate the vial at 60 °C overnight to dissolve the paraffin wax.
11. Replace the solution with paraffin wax as follows: remove 1/3 of the solution and add 1/3 of fresh paraffin wax melted at 60 ° C. Repeat this process three times a day for 3 days to ensure complete substitution of the Histo-Clear II with paraffin wax.
12. Place two plastic dishes containing melted paraffin wax on a hot plate set at 60 °C.
13. Decant the content of the vial, including the samples, into a plastic dish from **step 12**.
14. Transfer the samples using tweezers to the other plastic dish containing paraffin wax from **step 12**.
15. Cool down the plastic dish containing the samples and allow the paraffin to solidify. Store the embedded samples at 4 °C.

16. Cut sections from the samples using a rotary microtome at a thickness of 8 μM.
17. Place distilled water on a MAS adhesive slide and float the sections onto it.
18. Warm the slide at 42 °C on a hot plate to stretch the sections. Remove any excess distilled water and allow the samples to dry. The slide can be stored at 4 °C for 1 month.

3.3 Paraffin Removal, Re-hydration, and Blocking

1. Place the slide in a staining rack.
2. Immerse the staining rack in Histo-Clear II for 10 min (*see* **Note 5**).
3. Repeat **step 2** with fresh Histo-Clear II.
4. Immerse the staining rack in 1:1 Histo-Clear II/ethanol for 10 min.
5. Immerse the staining rack in the graded ethanol series (100%, 95%, 90%, 85%, 70%, 50%, 30%), followed by distilled water and PBS. Incubate in each solution for 1 min.
6. Remove the slide from the staining rack and wipe the back of the slide with a paper towel.
7. Apply 600 μL of proteinase K solution onto the slide, ensuring that the solution is evenly distributed across all sections.
8. Incubate at room temperature for 10 min.
9. Discard the proteinase K solution and place the slide to the staining rack.
10. Immerse in PBS for 5 min.
11. Repeat **step 10** with fresh PBS twice.
12. Place the slide in the humidity chamber.
13. Apply 600 μL of blocking solution onto the slide, ensuring that the solution is evenly distributed across all sections.
14. Incubate at room temperature for 30 min.
15. Discard the blocking solution and proceed to Subheading 3, **steps 4** or **5**.

3.4 FITC Staining

1. Prepare 50 μL of primary antibody solution per slide by diluting the OSH1 antibody from **step 42** in Subheading 3.1 in the blocking solution at a ratio of 1:5.
2. Place the slide in the humidity chamber.
3. Apply 50 μL of primary antibody solution from **step 1** onto the slide (*see* **Note 6**) and place a coverslip over it.
4. Incubate at room temperature overnight.
5. Immerse the slide in PBS-T and slip down the coverslip in PBS-T.

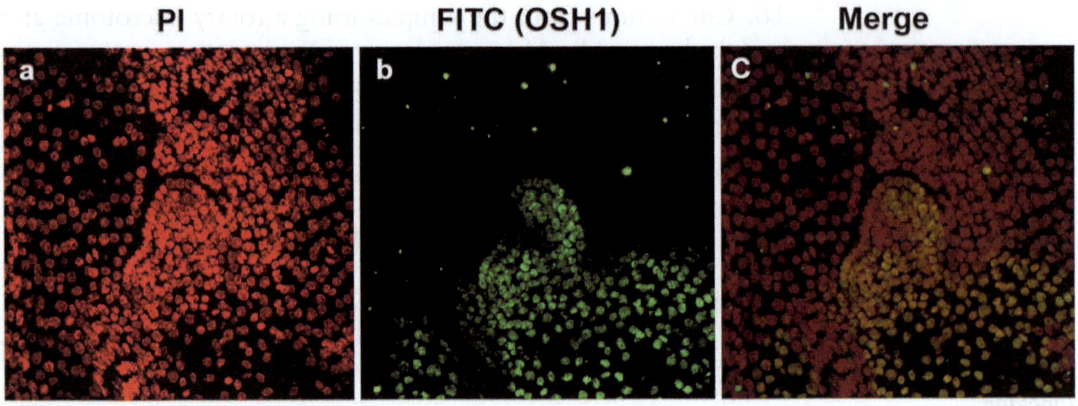

Fig. 2 Detection of OSH1 protein in rice SAM by IHC. (**a**) PI staining of rice SAM. (**b**) Detection of OSH1 protein by FITC staining in the same sections as (**a**). (**c**) Merged image of (**a**) and (**b**)

6. Place the slides to the staining rack and immerse in PBS-T for 15 min.
7. Immerse in fresh PBS-T for 15 min.
8. Prepare 100 μL of secondary antibody solution per slide by diluting the secondary antibody in the blocking solution at a ratio of 1:600.
9. Place the slide in the humidity chamber.
10. Apply 100 μL of secondary antibody from **step** 7 onto the slide and place a coverslip over it.
11. Incubate at room temperature for 2 h
12. Repeat **steps 5–7**.
13. Immerse in PBS for 2 min.
14. Apply 60 μL VECTASHIELD on the slide and place a coverslip over it. Observe using a confocal laser scanning microscope (*see* **Notes 7** and **8** and Fig. 2). Fluorescence fades after prolonged storage, so it is recommended to observe the fluorescence as soon as possible on the same day.

3.5 AP Staining

1. Prepare 100 μL of primary antibody solution per slide by diluting the OSH1 antibody from **step 42** in Subheading 3.1 in the blocking solution at a ratio of 1:200.
2. Place the slide in the humidity chamber.
3. Apply 100 μL of primary antibody solution from **step 1** onto the slide (*see* **Note 6**) and place a coverslip over it.
4. Incubate at room temperature overnight.
5. Immerse the slide in the blocking solution and slip down the coverslip in the blocking solution.
6. Place the slide in the humidity chamber.

7. Apply 600 μL of blocking solution onto the slide, ensuring that the solution is evenly distributed across all sections.
8. Incubate at room temperature for 15 min.
9. Discard the blocking solution.
10. Repeat **steps 7–9**.
11. Prepare 100 μL of secondary antibody solution per slide by diluting the secondary antibody in the blocking solution at a ratio of 1:600.
12. Place the slide in the humidity chamber.
13. Apply 100 μL of secondary antibody from **step 11** onto the slide and place a coverslip over it.
14. Incubate at room temperature for 2 h.
15. Repeat **steps 5–10**.
16. Place the slides to the staining rack and immerse in PBS-T for 2 min.
17. Immerse in TNM for 5 min.
18. Repeat **step 17** with fresh TNM.
19. Place the slide in the humidity chamber.
20. Apply 400 μL of AP substrate onto the slide and place a coverslip over it.
21. Incubate at room temperature in the dark to develop color (*see* **Note 9**).
22. Immerse the slide in TE and slip down the coverslip in TE.
23. Place the slides to the staining rack.
24. Immerse the staining rack in the graded ethanol series (30%, 60%, 80%, 95%, 100% ethanol), followed by 1:1 Lemosol/ ethanol and 100% Lemosol. Incubate in each solution for 1 min.
25. Apply 400 μL EUKITT mounting medium on the slide and place a coverslip over it. Observe using an optical microscope.

4 Notes

1. Either purchase a pre-cast 10% SDS-PAGE gel or prepare one yourself (*see* Chap. 5). Acrylamide concentrations should be 10% for the separating gel and 5% for the stacking gel.
2. OSH1 protein is included in the insoluble fraction in this step.
3. We conduct electrophoresis on 4 gels, each applied with about 400 μg of protein divided into 12 wells. For OSH1, the region around 40 kD is excised from each gel, and the protein is transferred from four-gel onto one PVDF membrane. We cut

out the necessary region of the gel, so blocking is not performed, but additional blocking may be necessary depending on the target protein.

4. We utilize an aspirator to decompress, gradually reducing the pressure inside the desiccator to −0.1 MPa for 10 min or longer.
5. If a slide rack is not available, fill the staining jar with the solution, ensuring that the slides are completely immersed. Agitate the solution in the staining jar by moving the slide rack up and down five times. Allow the slides to stand in the staining jar for the specified time.
6. The dilution ratio of the primary antibody should be titrated using known positive samples. A negative control should also be prepared where no primary antibody is added.
7. We use a confocal microscope BX61 (OLYMPUS). Select the appropriate filter for the excitation ($\lambda_{ex} = 495$ nm) and fluorescence ($\lambda_{em} = 525$ nm) of the FITC.
8. For simultaneous PI staining of the slide, apply 600 μL of 5 μg/ml RNase A in PBS onto the slides and incubate at 25 °C for 5 min to degrade RNA before mount. Discard RNase A solution and apply 600 μL of 0.1 μg/mL PI on the slide. Incubate at 25 °C for 5 min. Wash twice with PBS for 5 min and mount by VECTASHIELD.
9. Monitor the color development and stop the reaction when a clear contrast between the stained and unstained area is visible, which can take anywhere from 3 h to overnight.

References

1. Coons AH, Creech HJ, Jones RN (1941) Immunological properties of an antibody containing a fluorescence group. Proc Soc Exp Biol Med 47:200–202
2. Matos LL, Trufelli DC, Matos MGL et al (2010) Immunohistochemistry as an important tool in biomarkers detection and clinical practice. Biomark Insights 2010:9–20
3. Ramos-Vara JA (2005) Technical aspects of immunohistochemistry. Vet Pathol 42:405–426
4. Yun H, Lathrop KL, Leger AJ (2021) A whole-mount immunohistochemistry protocol for detection of mouse corneal nerves. STAR Protoc 2:100734
5. Tuson JR, Pascoe EW, Jacob DA (1990) A novel immunohistochemical technique for demonstration of specific binding of human monoclonal antibodies to human cryostat tissue sections. J Histochem Cytochem 38:923–926
6. Hira VVV, Jong AL, Ferro K et al (2019) Comparison of different methodologies and cryostat versus paraffin sections for chromogenic immunohistochemistry. Acta Histochem 121:125–134
7. Jackson D, Veit B, Hake S (1994) Expression of maize KNOTTED1 related homeobox genes in the shoot apical meristem predicts patterns of morphogenesis in the vegetative shoot. Development 120:405–413
8. Sato Y, Hong SK, Tagiri A et al (1996) A rice homeobox gene, OSH1, is expressed before organ differentiation in a specific region during early embryogenesis. Proc Natl Acad Sci USA 93:8117–8122
9. Sentoku N, Sato Y, Kurata N et al (1999) Regional expression of the rice KN1-type

homeobox gene family during embryo, shoot, and flower development. Plant Cell 11:1651–1663

10. Tsuda K, Ito Y, Sato Y et al (2011) Positive autoregulation of a *KNOX* gene is essential for shoot apical meristem maintenance in rice. Plant Cell 23:4368–4381

11. Nosaka-Takahashi M, Kato M, Kumamaru T et al (2022) Measurements of the number of specified and unspecified cells in the shoot apical meristem during a plastochron in rice (Oryza sativa) reveal the robustness of cellular specification process in plant development. PLoS One 17:1–17

Chapter 9

Agrobacterium-Mediated Transformation and Targeted Mutagenesis Using SpCas12f in Rice

Satoru Sukegawa, Seiichi Toki, and Hiroaki Saika

Abstract

Clustered regularly interspaced short palindromic repeats (CRISPR) and CRISPR-associated proteins (Cas) (CRISPR-Cas) is an adaptive prokaryote immune system against foreign DNA/RNA that is now applied widely to genome editing. A miniature Cas, CRISPR-Cas12f, is one-half to one-third the size of the CRISPR-Cas9 that is commonly used in genome editing experiments in many organisms, including higher plants. The compactness of CRISPR-Cas12f is expected to be advantageous in terms of vector construction and transformation frequency. Moreover, CRISPR-Cas12f can be useful for virus vector-mediated genome editing because the size of the transgene is the major restriction in the use of virus vectors. Here, we describe our protocol for targeted mutagenesis using Cas12f derived from *Syntrophomonas palmitatica* (SpCas12f) via *Agrobacterium*-mediated transformation in rice. We also summarize some approaches to improve the frequency of targeted mutagenesis using SpCas12f.

Key words *Agrobacterium*-mediated transformation, CRISPR-Cas12f, Genome editing, SpCas12f, Targeted mutagenesis

1 Introduction

Targeted mutagenesis—genome editing using sequence-specific nucleases—is a powerful tool with which to modify a targeted gene in both basic and applied research. The CRISPR-Cas system (clustered regularly interspaced short palindromic repeats (CRISPR)-CRISPR-associated proteins (Cas)) is an adaptive immune system in bacteria that was first reported as a sequence-specific nuclease and genome editing tool in 2012 [1]. CRISPR-Cas has since been widely applied to genome editing in many organisms, including mammalian cells and plants.

In rice, reliable transformation via *Agrobacterium* was reported 30 years ago [2], and an efficient *Agrobacterium*-mediated transformation system is now established in many rice varieties

Taiji Kawakatsu (ed.), *Rice: Methods and Protocols*, Methods in Molecular Biology, vol. 2869,
https://doi.org/10.1007/978-1-0716-4204-7_9,

[3–5]. *Agrobacterium*-mediated transformation is the main method used for targeted mutagenesis in rice when using the CRISPR-Cas9 system.

CRISPR-Cas is categorized into two classes: Class 2 CRISPR-Cas consists of a single protein and guide RNA (gRNA) such as Cas9, whereas Class 1 CRISPR-Cas consists of multiple proteins plus a gRNA. CRISPR-Cas9 (Cas9 derived from *Streptococcus pyogenes* (SpCas9), especially) is the most widely used Cas as a sequence-specific nuclease. Because CRISPR-Cas has different enzymatic properties, various kinds of CRISPR-Cas other than Cas9 have been developed as genome editing tools. CRISPR-Cas12f (originally named Cas14a) [6] belongs to type V-F CRISPR-Cas. Cas12f is a small protein (400–700 amino acids) that recognizes a T-rich protospacer adjacent motif (PAM) sequence. CRISPR-Cas12f cleaves double-stranded DNA to generate sticky ends [7, 8], although it was reported initially to cleave single-stranded DNA [6]. Cas12f forms an asymmetric dimer with one single gRNA (sgRNA) [9–12]. Each of the RuvC nuclease domains in Cas12f monomers forms an asymmetric interface, whereas each of the nucleic acid recognition lobe (REC) domains forms a symmetrical interface. This dimer recognizes the target sequence by forming a complex with sgRNA, and a single RuvC domain cleaves both strands of the target double-stranded DNA.

Cas12f derived from *Syntrophomonas palmitatica* (SpCas12f) has been applied to genome editing in bacteria, human cells, rice, maize, and tobacco [7, 13–15]. SpCas12f recognizes 5′-TTC-3′ as a PAM sequence and cleaves target DNA between 22 and 24 bp downstream of the PAM (*see* Fig. 1 and Table 1) [7]. In vitro analysis showed that the optimal conditions for DNA cleavage

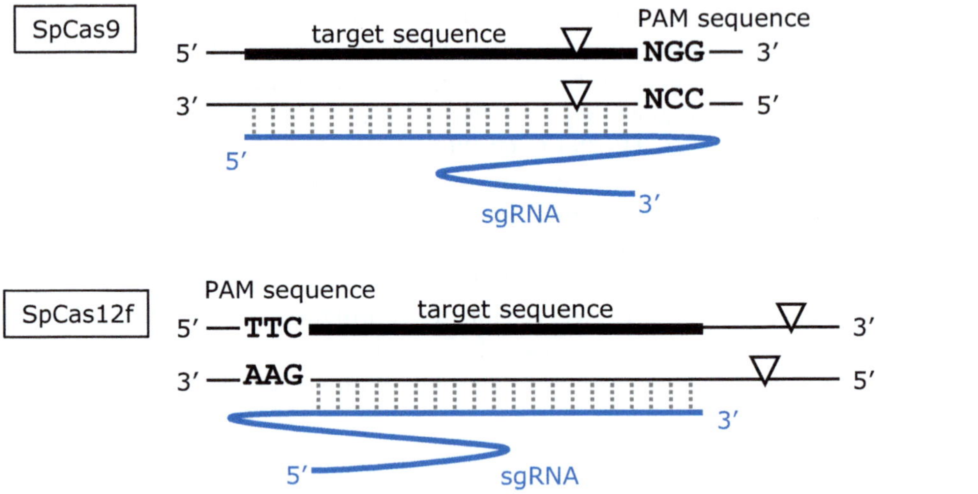

Fig. 1 Scheme of DNA cleavage by SpCas12f and SpCas9. Black and blue lines indicate target DNA and sgRNA, respectively. Triangles show cleavage sites

Table 1
Comparison of SpCas12f and SpCas9

	SpCas12f	SpCas9
Classification	Class 2 type V-F	Class 2 type II
Organism	*Syntrophomonas palmitatica*	*Streptococcus pyogenes*
Protein size	497 aa	1368 aa
PAM	5′-TTC-3′	5′-NGG-3′
DNA ends after cleavage	Sticky	Blunt
Formation	Dimer + one sgRNA	Monomer + one sgRNA

activity by SpCas12f are 45–55 °C and 200–250 mM NaCl [7]. In the case of maize, targeted mutagenesis was achieved by heat pulse treatment (4 h heat shock of 45 °C per day for 3 days) after biolistic transformation [7]. On the other hand, mutations were introduced successfully into target genes without any heat shock treatment in rice [14]. The main mutation pattern introduced by SpCas12f is deletion of several base pairs (bp) up to dozens of bp [7, 13–15].

The protocol described in this chapter covers vector construction for SpCas12f-mediated targeted mutagenesis, *Agrobacterium*-mediated transformation in rice, and electrophoretic detection and determination of mutations introduced by SpCas12f.

2 Materials

2.1 Vector Construction

1. Oligonucleotides for sgRNA: Select a 20-nt downstream PAM sequence as a target sequence. Prepare two oligonucleotides as follows (*see* Fig. 2 and **Note 1**):

   sgRNA forward oligo: 5′-AAACNNNNNNNNNNNNNNNNNNNNN-3′ (N = 20-nt target sequence).

   sgRNA reverse oligo: 5′-AAAANNNNNNNNNNNNNNNNNNNN-3′ (N = 20-nt reverse and complementary target sequence).
2. Guide RNA cloning vector: pUCAP-sgRNA containing the OsU6 promoter and the sgRNA scaffold (*see* Fig. 2) [14].
3. Binary vector for expression of SpCas12f: pPZP-SpCas12f containing the rice-codon-optimized SpCas12f expression cassette (*see* Fig. 2) [14].
4. Restriction enzymes: *Asc*I, *Pac*I, *Bbs*I (New England Biolabs).
5. Agarose gel electrophoresis system: Mupid One Electrophoresis System (Nippon Genetics), TAE buffer (dilute 50× TAE (NIPPONGENE) with H_2O), 1% (w/v) agarose gel in TAE buffer, loading dye, DNA size marker.

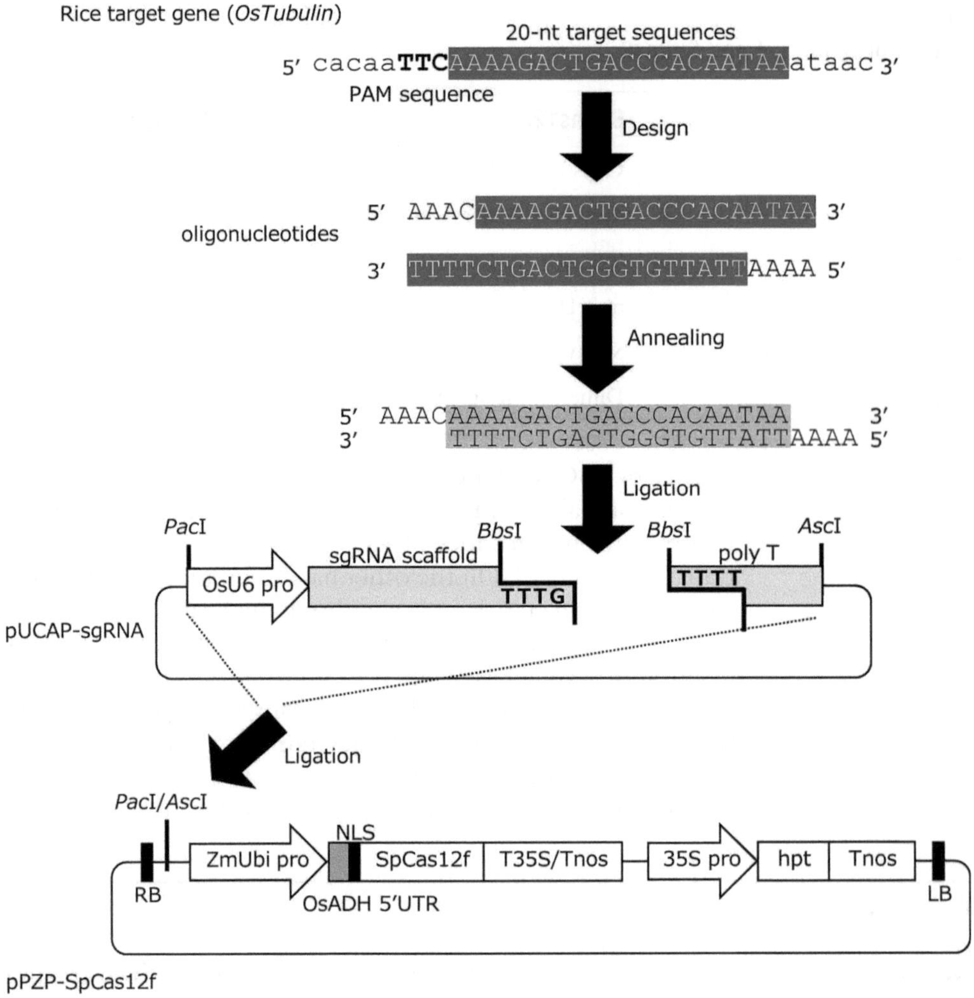

Fig. 2 Schematic representation of vector construction. An example of the construction of an SpCas12f vector targeting the OsTub-2 site [14]. Bold letters and white letters highlighted with gray in *OsTubulin* sequences are the PAM sequence and 20-nt target sequence, respectively. OsU6 pro, ZmUbi pro, and 35S pro: rice *U6–2* promoter, maize *Ubiquitin 1* promoter, and cauliflower mosaic virus 35S RNA promoter, respectively. OsADH 5′-UTR, the 5′ of an untranslated region of rice *alcohol dehydrogenase 2* gene as a translational enhancer; NLS, the nuclear localization signal derived from Simian virus 40; hpt, hygromycin phosphotransferase; T35S and TNos, the terminator region of cauliflower mosaic virus 35S RNA and nopaline synthase, respectively. LB and RB, left and right border sequences, respectively

6. Gel extraction kit.
7. Ligation kit: LigaFast Rapid DNA Ligation System (Promega). This kit includes 2× ligation buffer and T4 DNA ligase.
8. *Escherichia coli* DH5α competent cells.
9. LB medium: (Liquid) 1 bag/100 mL LB broth (Fujifilm). Autoclave at 121 °C for 20 min. After cooling, add 100 mg/L ampicillin or spectinomycin in a clean bench to keep the

media sterile. (Solid) 1 bag/100 mL LB agar (Fujifilm). Autoclave at 121 °C for 20 min. After cooling to approximately 60 °C, add 100 mg/L ampicillin or spectinomycin and dispense 20 mL of the medium into empty sterile Petri dishes (9 cm in diameter, 1.5 cm in depth) in a clean bench to keep the medium sterile. Store sterile media at 4 °C for up to 1 month.

10. Incubator for *E. coli* cultivation (37 °C).
11. Plasmid miniprep kit.
12. Reagents for Sanger sequencing: BigDye Terminator v3.1 cycle sequencing kit, BigDye Xterminator Purification Kit (Thermo Fisher Scientific).
13. Primers for sequencing:

    M13 forward: 5′-GTAAAACGACGGCCAGT-3′.

    M13 reverse: 5′-CAGGAAACAGCTATGAC-3′.
14. PCR thermal cycler.
15. Instruments for Sanger sequencing by capillary electrophoresis.

2.2 Agrobacterium-Mediated Transformation

1. Mature rice seeds (*Oryza sativa* L. cv. Nipponbare) (*see* **Note 2**).
2. *Agrobacterium tumefaciens* EHA105.
3. Electroporation device: *E. coli* Pulser Transformation Apparatus (Bio-Rad) and 0.1 cm cuvettes (Bio-Rad).
4. YEB buffer: 5 g/L bacto-peptone, 5 g/L beef extract, 1 g/L yeast extract, 5 g/L sucrose, 492.95 mg/L $MgSO_4 \cdot 7H_2O$. Adjust pH to 7.2 with NaOH and autoclave at 121 °C for 20 min. Store sterile media at −20 °C for up to 1 year.
5. LB solid/liquid medium containing 100 mg/L spectinomycin (*see* **item 9** in Subheading 2.1).
6. 80% glycerol.
7. 70% ethanol.
8. 2.5% sodium hypochlorite: Mix 20 mL sodium hypochlorite solution (Fujifilm) and 20 mL sterilized water.
9. 2.5% sodium hypochlorite containing 0.1% Tween 20: Add 40 μL of polyoxyethylene [20] sorbitan monolaurate (Tween 20, Fujifilm) to 40 mL of 2.5% sodium hypochlorite.
10. N6 vitamin (1000×): 0.5 mg/mL nicotinic acid, 0.5 mg/mL pyridoxine HCl, 1 mg/mL thiamine HCl.
11. MS vitamin (1000×): 0.5 mg/mL nicotinic acid, 0.5 mg/mL pyridoxine HCl, 0.1 mg/mL thiamine HCl.

12. N6D medium: 30 g/L sucrose, 0.3 g/L casamino acids, 1 bag/L Chu (N6) medium salt mixture (Fujifilm), 2.878 g/L L-proline, 2 mg/L L-glycine, 100 mg/L *myo*-inositol, 1 mL/L N6 vitamin (1000×), 2 mg/L 2,4-D, and 4 g/L Gelrite. Adjust pH to 5.8 with KOH and autoclave at 121 °C for 20 min. After cooling to approximately 60 °C, dispense 50 mL of the medium into empty sterile Petri dishes (9 cm in diameter, 2 cm in depth) in a clean bench to keep the medium sterile. For the selection of transformed calli, add 50 mg/L hygromycin and 25 mg/L meropenem (Fujifilm) into cooled media before dispensing. Store sterile media at 4 °C for up to 2 months.
13. AB medium: 5 g/L glucose, 3 g/L K_2HPO_4, 1.3 g/L $NaH_2PO_4 \cdot 2H_2O$, 1 g/L NH_4Cl, 150 mg/L KCl, 10 mg/L $CaCl_2 \cdot 2H_2O$, 2.5 mg/L $FeSO_4 \cdot 7H_2O$, and 15 g/L Bacto Agar. Adjust pH to 7.2 with NaOH and autoclave at 121 °C for 20 min. After cooling to approximately 60 °C, add 1.2 mL/L of sterile 1 M $MgSO_4 \cdot 7H_2O$ and 100 mg/L spectinomycin and 25 mg/L rifampicin and dispense 20 mL of the medium into empty sterile Petri dishes (9 cm in diameter, 1.5 cm in depth) in a clean bench to keep the medium sterile. Store sterile media at 4 °C for up to 1 month.
14. AAM medium: 68.5 g/L sucrose, 36 g/L glucose, 250 mg/L $MgSO_4 \cdot 7H_2O$, 150 mg/L $CaCl_2 \cdot 2H_2O$, 150 mg/L $NaH_2PO_4 \cdot 2H_2O$, 3 g/L KCl, 40 mg/L Fe-EDTA, 10 mg/L $MnSO_4 \cdot 4H_2O$, 2 mg/L $ZnSO_4 \cdot 7H_2O$, 0.025 mg/L $CuSO_4 \cdot 5H_2O$, 0.025 mg/L $CoCl_2 \cdot 6H_2O$, 0.75 mg/L KI, 3 mg/L H_3BO_3, 0.25 mg/L $Na_2MoO_4 \cdot 2H_2O$, 100 mg/L *myo*-inositol, 1 mg/L nicotinic acid, 1 mg/L pyridoxine HCl, 10 mg/L thiamine HCl, 0.5 g/L casamino acids, 7.5 mg/L L-glycine, 176.7 mg/L L-arginine, 0.9 g/L L-glutamine, and 0.3 g/L L-aspartic acid. Adjust pH to 5.2 with KOH and autoclave at 121 °C for 20 min. Just before use, add 20 mg/L acetosyringone (3′,5′-dimethoxy-4′-hydroxyacetophenone). Store sterile media at 4 °C for up to 2 months.
15. 2N6AS medium: 30 g/L sucrose, 10 g/L glucose, 0.3 g/L casamino acids, 1 bag/L Chu (N6) medium salt mixture (Fujifilm), 2 mg/L L-glycine, 100 mg/L *myo*-inositol, 1 mL/L N6 vitamin (1000×), 2 mg/L 2,4-D, and 4 g/L Gelrite. Adjust pH to 5.2 with KOH and autoclave at 121 °C for 20 min. After cooling to approximately 60 °C, add 20 mg/L acetosyringone and dispense 40 mL of the medium into empty sterile Petri dishes (9 cm in diameter, 2 cm in depth) in a clean bench to keep the medium sterile. Store sterile media at 4 °C for up to 2 months.

16. ReIII medium: 30 g/L sucrose, 30 g/L sorbitol, 2 g/L casamino acids, 1 bag/L Murashige and Skoog plant salt mixture (Fujifilm), 2 mg/L L-glycine, 100 mg/L *myo*-inositol, 1 mL/L MS vitamin (1000×), 20 μg/L NAA, 2 mg/L kinetin, and 4 g/L Gelrite. Adjust pH to 5.8 with KOH and autoclave at 121 °C for 20 min. After cooling to approximately 60 °C, add 25 mg/L meropenem and dispense 50 mL of the medium into empty sterile Petri dishes (9 cm in diameter, 2 cm in depth) in a clean bench to keep the medium sterile. Store sterile media at 4 °C for up to 2 months.
17. Hormone-free medium: 30 g/L sucrose, 1 bag/L Murashige and Skoog plant salt mixture (Fujifilm), 2 mg/L L-glycine, 100 mg/L *myo*-inositol, 1 mL/L MS vitamin (1000×), and 4 g/L Gelrite. Adjust pH to 5.8 with KOH and autoclave at 121 °C for 20 min. After cooling to approximately 60 °C, add 25 mg/L meropenem and dispense 50 mL of the medium into empty sterile Petri dishes (9 cm in diameter, 2 cm in depth) in a clean bench to keep the medium sterile. Store sterile media at 4 °C for up to 2 months.

2.3 DNA Extraction and Mutation Analysis

1. Agencourt Chloropure (Beckman Coulter).
2. KOD One PCR Master Mix (TOYOBO).
3. MultiNA microchip electrophoresis device (Shimadzu) (optional; *see* **Note 3**).
4. Zero Blunt TOPO PCR cloning kit (Thermo Fisher Scientific).
5. *E. coli* DH5α competent cells.
6. LB solid medium containing 100 mg/L kanamycin (*see* **item 9** in Subheading 2.1).
7. Incubator for *E. coli* cultivation (37 °C).
8. Target-specific primers: Design PCR primers to amplify 200–300 bp around the target site of SpCas12f.
9. M13 primers (*see* **item 13** in Subheading 2.1).
10. Reagents for Sanger sequencing (*see* **item 12** in Subheading 2.1).
11. PCR thermal cycler.
12. Instruments for Sanger sequencing by capillary electrophoresis.

3 Methods

3.1 Vector Construction

The outline of the vector construction is shown in Fig. 2 and **Note 4**.

3.2 Rice Transformation

An outline of the rice transformation protocol is shown in Fig. 3. Perform all steps in this subheading except for **step 1** on a clean bench to prevent contamination.

1. Transform the binary vector produced in Subheading 3.1 into *Agrobacterium* EHA105 competent cells by electroporation (1.8 kV, 5 ms). After electroporation, add 10 times the volume of YEB buffer and culture at 28 °C for 1–2 h with shaking (180 rpm). Culture and select transformed *Agrobacterium* EHA105 on LB solid medium containing 100 mg/L spectinomycin for 3 days at 28 °C. Pick a colony and culture it with LB liquid medium containing 100 mg/L spectinomycin at 28 °C with shaking (180 rpm) overnight. Mix 750 μL of *Agrobacterium* EHA105 liquid culture and 250 μL sterile 80% glycerol and store at −80 °C.
2. Wash and sterilize the surface of dehulled rice seeds with 70% ethanol for 30 s, 2.5% sodium hypochlorite solution containing 0.1% Tween 20 for 15 min, and 2.5% sodium hypochlorite solution for 15 min. After each step, discard the solution and wash the seeds with sterile water several times.
3. Place sterilized seeds on N6D medium and culture at 30–33 °C for 3–4 weeks under constant light. Prepare a few plates with 16 seeds per plate.
4. Transfer emerging secondary calli to fresh N6D medium and culture at 30–33 °C for 3–5 days under constant light.
5. Streak *Agrobacterium* EHA105 containing the binary vector from glycerol stock on AB medium containing 100 mg/L spectinomycin and 25 mg/L rifampicin and culture at 28 °C for 3 days in the dark.
6. Suspend one loop of cultured *Agrobacterium* EHA105 thoroughly in 40 mL of AAM medium supplemented with acetosyringone.
7. Immerse 2.5–5 mL of rice secondary calli in *Agrobacterium* solution from **step 6** for 1.5 min.
8. Discard the solution and place inoculated calli on 2N6AS medium and culture at 23 °C in the dark for 3 days.
9. Wash infected calli with sterilized water containing 25 mg/L meropenem 7–8 times.
10. Place washed calli on N6D medium containing 50 mg/L hygromycin and 25 mg/L meropenem (4–5 plates with 50–100 pieces of calli per plate) and culture at 30–33 °C under constant light.
11. At 2 weeks after onset of selection, transfer all pieces of calli except for clearly dead and browning calli to fresh N6D medium containing 50 mg/L hygromycin and 25 mg/L meropenem (4–5 plates with 50–100 pieces of calli per plate).

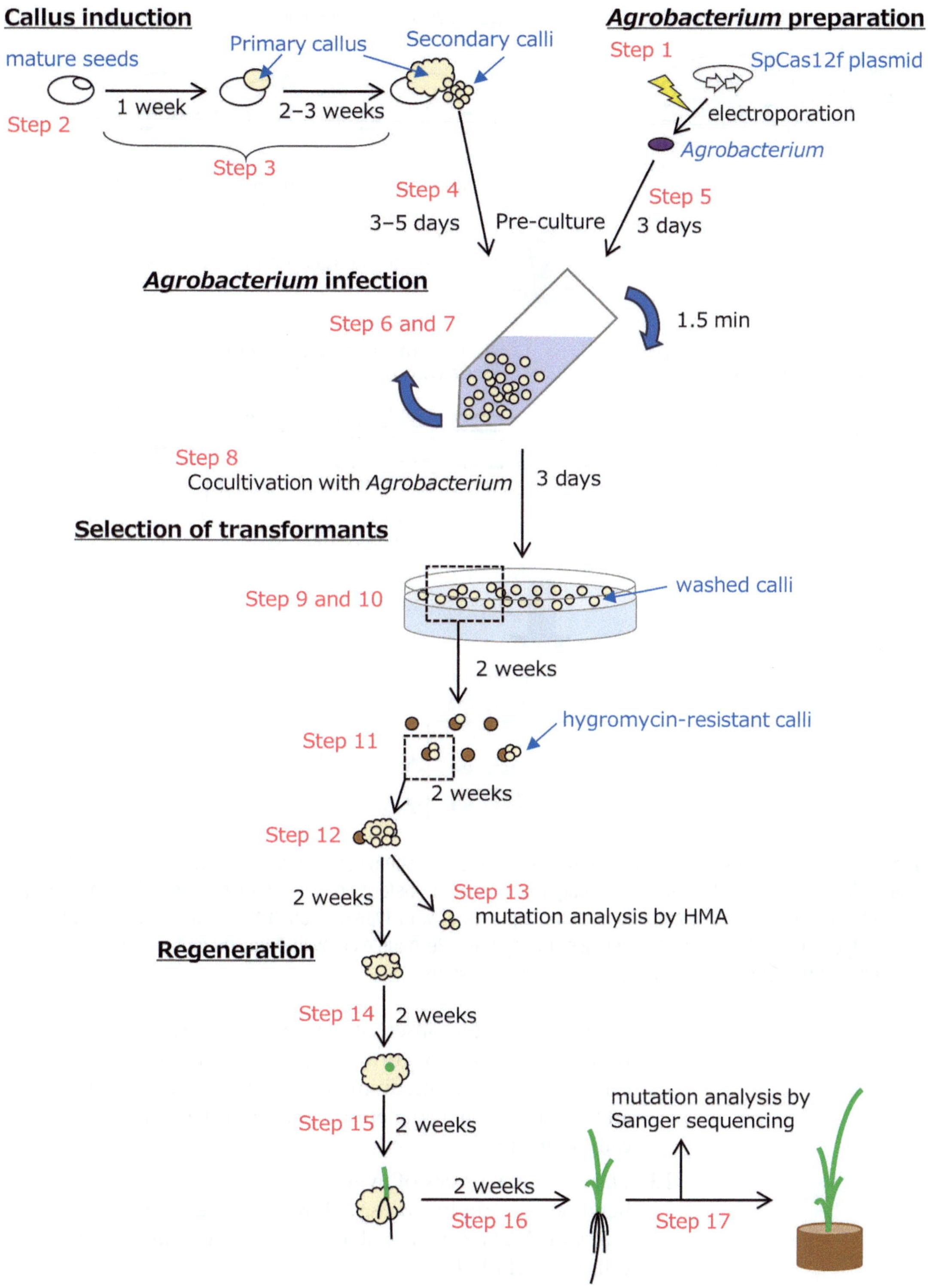

Fig. 3 Schematic representation of rice transformation and targeted mutagenesis

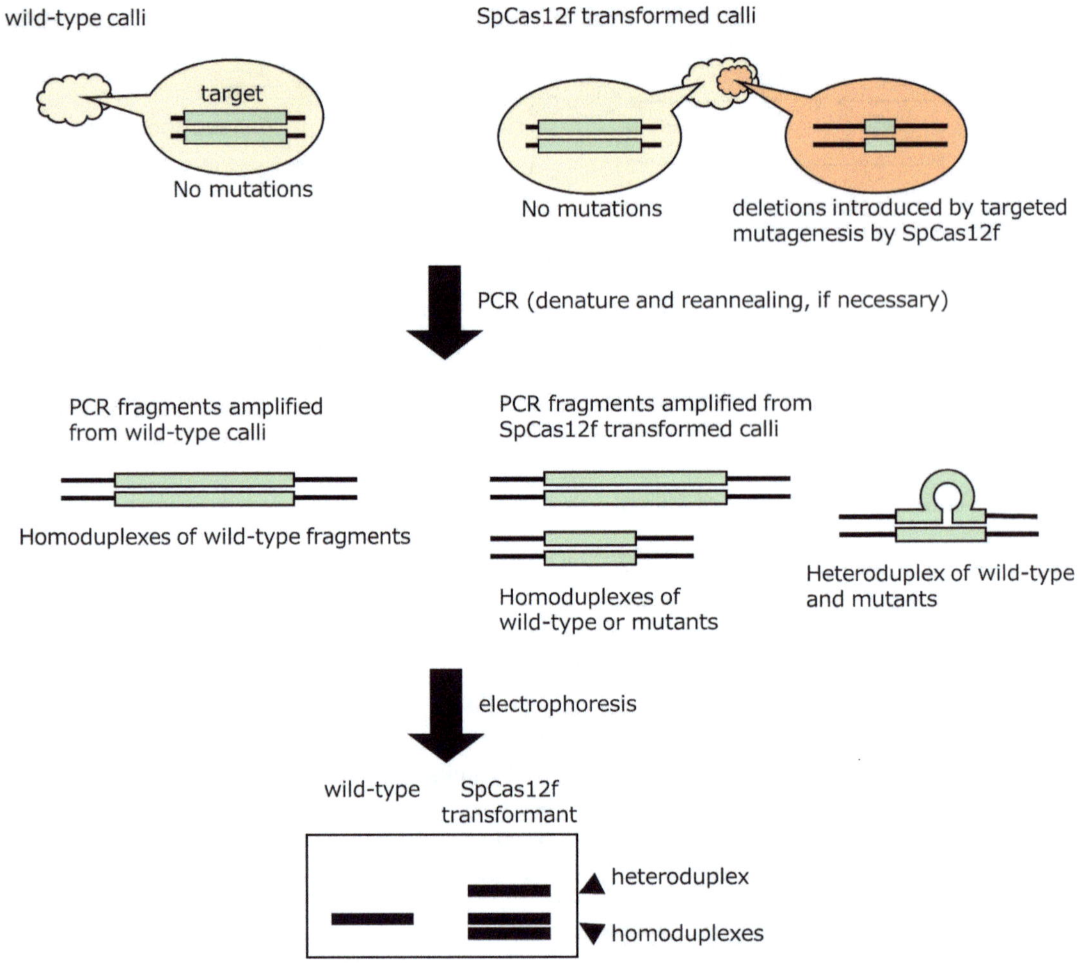

Fig. 4 Principles of HMA. An example of transformed calli in some of which deletion has occurred at the target site by SpCas12f-mediated targeted mutagenesis. Since transformed calli are mosaics of wild-type cells and mutant cells, three types of amplicons, homoduplexes of wild-type or mutant fragments, and heteroduplexes of wild-type and mutant fragments are generated. The heteroduplex of wild-type and mutant fragments can be distinguished by electrophoresis because of mobility shifts

12. At 4 weeks after onset of selection (2 weeks after the first transfer), transfer vigorously growing hygromycin-resistant calli to fresh N6D medium containing 50 mg/L hygromycin and 25 mg/L meropenem (at least 5 plates with 25 pieces of calli per plate).
13. Harvest a few pieces of hygromycin-resistant calli and perform heteroduplex mobility assay (HMA; *see* Fig. 4 and Subheading 3.3, **step 1**). Select the calli in which a large number of mutant cells are included.
14. Transfer selected calli to ReIII medium (a few pieces of calli, ~9 lines per plate) and culture at 27 °C under 16 h light/8 h dark for 2 weeks.

15. Transfer calli to fresh ReIII medium after 2 weeks of onset of cultivation on ReIII medium.
16. Transfer regenerating plants to hormone-free medium and culture at 27 °C under 16 h light/8 h dark for 2 weeks. Confirm mutations by Sanger sequencing (*see* Subheading 3.3, **step 2**).
17. Transfer the rice seedlings to soil and grow in a greenhouse.

3.3 Detection of Mutations

3.3.1 Mutation Analysis by HMA

1. Harvest hygromycin-resistant calli after 5 weeks of onset of selection with hygromycin from **step 13** in Subheading 3.2. Extract genomic DNA using Agencourt Chloropure, following the manufacturer's protocols.
2. Amplify the region including the target site from genomic DNA extracted in **step 1**. Add the following components to a sterile nuclease-free PCR tube (*see* **Note 5**):

Genomic DNA	5 μL
10 μM target-specific forward primer	1.5 μL
10 μM target-specific reverse primer	1.5 μL
KOD one PCR master mix	25 μL
H_2O	17 μL
Total	50 μL

3. Run the PCR reaction as follows:
   2 min at 98 °C
   35 cycles of:
   10 s at 98 °C
   5 s at 60 °C
   5 s at 68 °C
   7 min at 68 °C
   hold at 12 °C
4. Resolve the PCR products (1/15 diluted) on a MultiNA microchip electrophoresis device, following the manufacturer's protocol. Find samples with extra bands other than a band corresponding to wild type (*see* **Notes 3**, **6** and 7). Estimate the mutation frequency (ratio of the number of mutated calli to that of SpCas12f-transformed calli).

3.3.2 Mutation Analysis by Sanger Sequencing

1. Harvest leaves of regenerated plants. Extract genomic DNA from leaves with a DNA extraction kit following the manufacturer's protocols.
2. Perform PCR amplification as **steps 2** and **3** in Subheading 3.3, **step 1**.

3. Confirm the mutation by Sanger sequencing with a target-specific primer. Add the following components to a sterile nuclease-free PCR tube (*see* **Note 5**):

PCR products (1/20 diluted; *see* **Note 8**)	2 μL
10 μM target-specific forward or reverse primer	0.32 μL
5× sequencing buffer	3 μL
BigDye terminator v3.1	2 μL
H_2O	12.68 μL
Total	20 μL

4. Run the cycle sequencing reaction as follows:
   2 min at 96 °C
   25 cycles of:

   10 s at 96 °C

   5 s at 50 °C

   4 min at 60 °C

   hold at 12 °C
5. Purify sequencing reaction products using the BigDye Xterminator Purification Kit following the manufacturer's protocol.
6. Perform sequencing with an instrument for Sanger sequencing by capillary electrophoresis following the manufacturer's protocol. Assemble and analyze the data with sequence analysis software (*see* **Note 9**).

Proceed with the steps below if the sequence of each PCR fragment needs to be confirmed by Sanger sequencing (optional):

7. Perform subcloning of PCR products from samples with mutation using Zero Blunt PCR cloning kit following the manufacturer's protocol.
8. Pick a single colony and perform colony PCR. Add the following components to a sterile nuclease-free PCR tube (*see* **Note 5**):

E. coli colony	–
10 μM M13 forward primer	1.5 μL
10 μM M13 reverse primer	1.5 μL
KOD one PCR master mix	25 μL
H_2O	22 μL
Total	50 μL

9. Run the PCR reaction as follows:
   2 min at 98 °C
   25 cycles of:

   10 s at 98 °C

   5 s at 60 °C

   5 s at 68 °C

   7 min at 68 °C
   hold at 12 °C
10. Confirm the mutation by Sanger sequencing with M13 forward or reverse primer. Add the following components to a sterile nuclease-free PCR tube:

Colony PCR products (1/20 diluted; *see* **Note 8**)	2 μL
10 μM M13 forward or reverse primer	0.32 μL
5× sequencing buffer	3 μL
BigDye terminator v3.1	2 μL
H_2O	12.68 μL
Total	20 μL

11. Run the cycle sequencing reaction as follows:
    2 min at 96 °C
    25 cycles of:

    10 s at 96 °C

    5 s at 50 °C

    4 min at 60 °C

    hold at 12 °C
12. Repeat **steps 5** and **6**.

4 Notes

1. Selection of the target sequence is thought to be one of the main factors determining the frequency of targeted mutagenesis. To date, no information on target sequence preference or target site prediction tool is available for SpCas12f-mediated targeted mutagenesis. However, a poly-T sequence "TTTT" should be avoided because it can be a termination signal for RNA polymerase III, which directs transcription of the sgRNA driven by the U6 promoter [16].
2. We have not tested other rice varieties, but we expect that SpCas12f-mediated targeted mutagenesis will work in rice varieties other than Nipponbare that can be transformed using

Agrobacterium, such as the japonica elite variety Koshihikari [17] and the model indica variety (aus subgroup) Kasalath [18]. A protocol for *Agrobacterium*-mediated transformation using immature embryos that can be applied to many varieties, including true indica varieties such as IR64, has been published [4].

3. MultiNA is a microchip electrophoresis system that can be used for DNA/RNA analysis. In our experience, mutations can be detected with HMA by electrophoresis of PCR products amplified with the KOD One PCR Master Mix without treatment on MultiNA. However, to reliably detect mutations, it is better to produce a heteroduplex of PCR fragments by denaturing and reannealing [19].

4. First, two oligonucleotides are synthesized. In the case of OsTub-2 [14], 5′-AAAC+AAAAGACTGACCCACAATAA-3′ and 5′-AAAA+TTATTGTGGGTCAGTCTTTT-3′ are prepared (*see* Fig. 2). Next, for a sgRNA expression cassette, digest pUCAP-sgRNA vector using *Bbs*I and purify the resulting fragments using a gel extraction kit. Denature and anneal the two oligonucleotides by the heat treatment. Ligate annealed oligonucleotides into the pUCAP-sgRNA vector. Transform the ligation products into *E. coli* competent cells and select transformed *E. coli* on LB solid medium containing ampicillin. Pick colonies and culture them separately in LB liquid medium containing ampicillin with vigorous shaking at 37 °C overnight. Extract plasmids using a plasmid miniprep kit and sequence them using M13 forward or reverse primer to confirm whether annealed oligonucleotides are inserted into the *Bbs*I site. Finally, digest the resultant sgRNA expression vector and pPZP-SpCas12f by *Asc*I/*Pac*I. After purification with a gel extraction kit, ligate the fragments obtained. Transform the ligation products into *E. coli* competent cells and select transformed *E. coli* on LB solid medium containing spectinomycin. Pick colonies and culture them separately in LB liquid medium containing spectinomycin with vigorous shaking at 37 °C overnight. Extract plasmids using a plasmid miniprep kit and perform a sequencing reaction using M13 forward primer to confirm that the sgRNA expression cassette is inserted into the *Asc*I/*Pac*I site.

5. The total volume of PCR and sequencing reaction can be reduced to 10 μL.

6. Since HMA is a method for the detection of heteroduplexes, mutations cannot be detected by HMA in clonally propagated calli harboring homozygous mutations (the same mutations in both alleles).

Table 2
Approaches for the enhancement of targeted mutagenesis frequency by SpCas12f

		Maize [7]	Rice [14]	*N. benthamiana* [15]
Approaches	Heat treatment	Positive (45 °C × 4 h/d × 3 d)	(not tested)	No effect (40 °C × 4 h/d × 4 d)
	18-nt spacer	(not tested)	Positive (unpublished data)	No effect
	Salt concentration	(not tested)	(not tested)	No effect
	sgRNA truncation [13]	(not tested)	(not tested)	(not tested)
Experimental condition	Transformation	Biolistics (stable transformation)	*Agrobacterium* (stable transformation)	Agroinfiltration (transient transformation) Geminivirus replicon
	Samples	Immature embryo 3 days after transformation Regenerated plants	Calli selected for 5–6 weeks Regenerated plants and their progenies	Leaves 3 days recovery and subsequent 4 days heat treatment

7. The frequency of SpCas12f-mediated targeted mutagenesis is lower than that of SpCas9. Some approaches to improve targeted mutagenesis frequency have been proposed (*see* Table 2). A pulse heat shock is necessary for SpCas12f-mediated targeted mutagenesis in maize [7]. The mutation frequencies using 18-nt guide sequences were enhanced by 2- to 4.5-fold compared to frequencies obtained using 20-nt guide sequences in rice (unpublished data). These approaches could potentially improve the mutagenesis frequency in rice, although they have been proven not to work in *Nicotiana benthamiana* [15].
8. Purification of PCR products is not necessary in this step.
9. Online applications to obtain mutated sequences from raw sequence data such as TIDE [20] and CRISP-ID [21] are available.

References

1. Jinek M, Chylinski K, Fonfara I et al (2012) A programmable dual-RNA-guided DNA endonuclease in adaptive bacterial immunity. Science 337:816–821
2. Hiei Y, Ohta S, Komari T et al (1994) Efficient transformation of rice (*Oryza sativa* L.) mediated by *Agrobacterium* and sequence analysis of the boundaries of the T-DNA. Plant J 6: 271–282
3. Nishimura A, Aichi I, Matsuoka M (2006) A protocol for *agrobacterium*-mediated transformation in rice. Nat Protoc 1:2796–2802
4. Hiei Y, Komari T (2008) *Agrobacterium*-mediated transformation of rice using immature embryos or calli induced from mature seed. Nat Protoc 3:824–834
5. Toki S, Hara N, Ono K et al (2006) Early infection of scutellum tissue with *Agrobacterium* allows high-speed transformation of rice. Plant J 47:969–976
6. Harrington LB, Burstein D, Chen JS et al (2018) Programmed DNA destruction by miniature CRISPR-Cas14 enzymes. Science 362: 839–842

7. Bigelyte G, Young JK, Karvelis T et al (2021) Miniature type V-F CRISPR-Cas nucleases enable targeted DNA modification in cells. Nat Commun 12:6191
8. Karvelis T, Bigelyte G, Young JK et al (2020) PAM recognition by miniature CRISPR-Cas12f nucleases triggers programmable double-stranded DNA target cleavage. Nucleic Acids Res 48:5016–5023
9. Takeda SN, Nakagawa R, Okazaki S et al (2021) Structure of the miniature type V-F CRISPR-Cas effector enzyme. Mol Cell 81: 558–570.e553
10. Hino T, Omura SN, Nakagawa R et al (2023) An AsCas12f-based compact genome-editing tool derived by deep mutational scanning and structural analysis. Cell 186:4920–4935.e4923
11. Xiao R, Li Z, Wang S et al (2021) Structural basis for substrate recognition and cleavage by the dimerization-dependent CRISPR-Cas12f nuclease. Nucleic Acids Res 49:4120–4128
12. Su M, Li F, Wang Y et al (2024) Molecular basis and engineering of miniature Cas12f with C-rich PAM specificity. Nat Chem Biol 20:180–189
13. Wang Y, Wang Y, Pan D et al (2022) Guide RNA engineering enables efficient CRISPR editing with a miniature Syntrophomonas palmitatica Cas12f1 nuclease. Cell Rep 40: 111418
14. Sukegawa S, Nureki O, Toki S et al (2023) Genome editing in rice mediated by miniature size Cas nuclease SpCas12f. Front Genome Ed 5:1138843
15. Gong Z, Previtera DA, Wang Y et al (2024) Geminiviral-induced genome editing using miniature CRISPR/Cas12j (CasΦ) and Cas12f variants in plants. Plant Cell Rep 43:71
16. Sekine R, Kawata T, Muramoto T (2018) CRISPR/Cas9 mediated targeting of multiple genes in Dictyostelium. Sci Rep 8:8471
17. Nishizawa-Yokoi A, Cermak T, Hoshino T et al (2016) A defect in DNA Ligase4 enhances the frequency of TALEN-mediated targeted mutagenesis in rice. Plant Physiol 170:653–666
18. Saika H, Toki S (2010) Mature seed-derived callus of the model *indica* rice variety Kasalath is highly competent in *Agrobacterium*-mediated transformation. Plant Cell Rep 29: 1351–1364
19. Kakui H, Yamazaki M, Shimizu KK (2021) PRIMA: a rapid and cost-effective genotyping method to detect single-nucleotide differences using probe-induced heteroduplexes. Sci Rep 11:20741
20. Brinkman EK, Chen T, Amendola M et al (2014) Easy quantitative assessment of genome editing by sequence trace decomposition. Nucleic Acids Res 42:e168–e168
21. Dehairs J, Talebi A, Cherifi Y et al (2016) CRISP-ID: decoding CRISPR mediated indels by Sanger sequencing. Sci Rep 6:28973

Chapter 10

Gene Targeting via CRISPR/Cas9-Mediated DNA Double-Strand Break Induction in Rice

Ayako Nishizawa-Yokoi and Seiichi Toki

Abstract

Gene targeting (GT) is a precise genome editing tool to achieve desired modification of a target gene, e.g., introduction of point mutations, knock-in of a reporter gene, or swapping of a functional domain, through homologous recombination. However, due to its low frequency, it has proved difficult to establish a universal GT system. The availability of the CRISPR/Cas9 system for genome editing in plants has opened up possibilities to apply GT successfully to some plant species. Here, we provide protocols for CRISPR/Cas9-mediated DNA double-strand break (DSB)-induced GT in rice.

Key words Gene targeting, CRISPR/Cas9, DNA double-strand breaks, Homologous recombination, Genome editing

1 Introduction

Homologous recombination (HR)-mediated gene targeting (GT) is a precise genome editing tool that enables pinpoint and/or large-scale alterations of an endogenous target gene, including the introduction of amino acid substitutions, induction of desired deletions or insertions, knock-in of a reporter gene or Flag-tag, and/or swapping of a functional domain. Unfortunately, the frequency of GT in higher plants is very low, because HR repair is thought to be a minor pathway in most multicellular organisms. To overcome this problem, many groups have made great efforts to establish universal GT, testing a large number of approaches in plants [1, 2]. In several organisms, including rice, positive–negative selection has been established to enrich cells harboring the GT-modified target gene and to eliminate transgenic cells carrying the transgene integrated randomly into other gene loci [3, 4]. Using this system, GT could potentially be applied to the precise genome editing of almost all endogenous genes throughout the genome; however, although this has been demonstrated successfully in rice, to date,

Taiji Kawakatsu (ed.), *Rice: Methods and Protocols*, Methods in Molecular Biology, vol. 2869,
https://doi.org/10.1007/978-1-0716-4204-7_10,

there has been no successful example of GT with positive–negative selection in other plant species.

A breakthrough in the improvement of GT frequency in plants was achieved with the use of sequence-specific endonucleases to induce double-strand breaks (DSBs) at the target locus. Since CRISPR/Cas9 has been available for plant genome editing, the CRISPR/Cas9-mediated DSB-induced GT system has been applied successfully to a number of plant species, including *Arabidopsis*, tomato, tobacco, wheat, and rice. In this chapter, we provide protocols for an efficient GT system in rice via a CRISPR/Cas9-mediated DSB induction system using an all-in-one vector comprised of CRISPR/Cas9 expression cassette, selection marker, and GT donor template carrying a sequence homologous to the target gene carrying the desired mutation(s) (*see* Fig. 1) [5]. While it

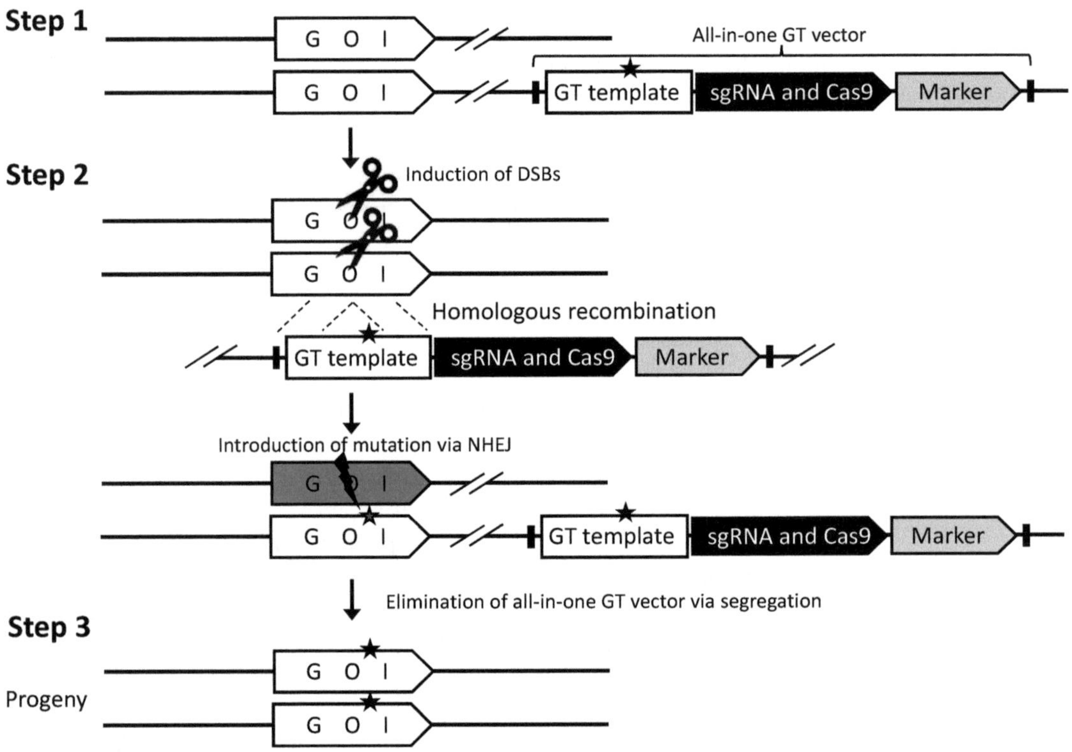

Fig. 1 Strategy for introduction of desired mutations into a gene of interest via CRISPR/Cas9-mediated DSB-induced GT in rice. (**Step 1**) All-in-one GT vector harboring CRISPR/Cas9 expression cassette, selection marker, and GT template carrying a homologous sequence of target genes with desired mutations (shown as a black star) transforms into rice calli via *Agrobacterium*. (**Step 2**) During the selection of transgenic calli on the medium containing antibiotics, sgRNA, and Cas9 express and induce the targeted mutation at the gene of interest (GOI). DSB-induced CRISPR/Cas9 is repaired preferentially by the non-homologous end joining (NHEJ) pathway, resulting in the induction of targeted mutations (e.g., small deletion or insertion) at the DSB site. DSBs are also repaired via HR using a GT template on the transgene integrated into the genome even though its frequency is very low. (**Step 3**) Transgene-free and homozygous plants with the modified GOI can be obtained via self-pollinating or out-crossing

remains challenging in almost all other plant species, this system is now available as a routine genome editing tool in rice.

2 Materials

2.1 Vector Construction

1. GT donor template: Double-strand DNA fragment consisting of three main parts: 5′ homology arm, sequences for modification, and 3′ homology arm. Homology arms are sequences that match regions upstream and downstream of the target site. Approximately 1 kb homology arms in total (0.5 kb upstream and 0.5 kb downstream sequences flanking the target site) are generally sufficient to obtain transgenic calli carrying the target gene modified by GT (*see* Fig. 2) [5]. Prepare GT donor template by gene synthesis, overlap extension PCR, or cloning (*see* **Note 1**).
2. sgRNA expression cassette: Double-strand DNA fragment containing the OsU6 promoter, the single guide RNA (sgRNA) target sequence, and the sgRNA scaffold. Design sgRNA at endogenous target sites between the 5′ and 3′ homology arms of the GT donor template to avoid digesting the GT donor template. Select a 20-nt sequence downstream PAM sequence as a target sequence. Cas9 recognizes 5′-NGG-3′ as a PAM sequence. Prepare at least two sgRNA expression cassettes as the use of multiple sgRNAs enhances the efficiency of CRISPR/Cas9-mediated GT, as demonstrated in

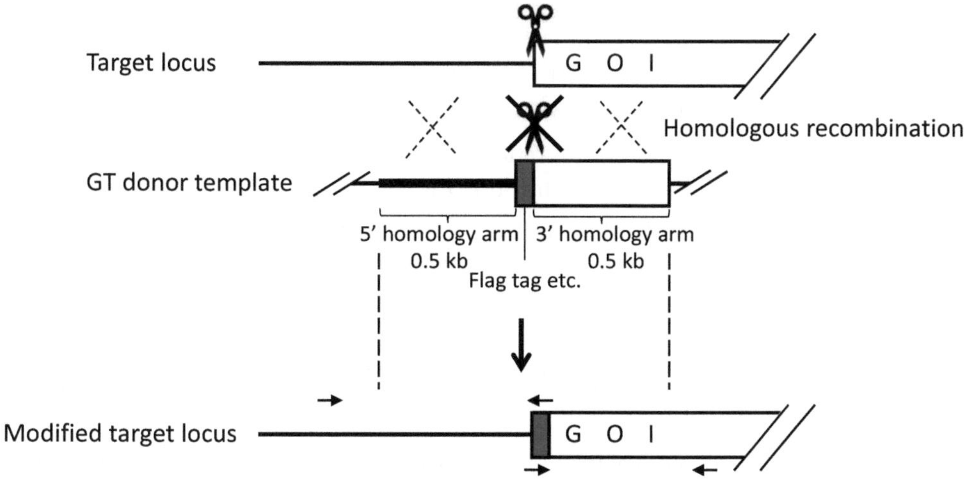

Fig. 2 Schematic diagram of the target locus (top), GT donor template (middle), and GT-modified target locus (bottom). White and gray boxes reveal the coding region of the GOI and desired modifications, respectively. Scissors show the target site of CRISPR/Cas9. Arrows shown in the upper and lower of the modified target locus reveal the location of the PCR primer sets for identifying the calli and plants carrying the GT-modified target

Arabidopsis [6]. In this chapter, we utilize two pU6gRNA plasmids as templates, each containing a designed sgRNA [7, 8]. Alternatively, synthesized DNA fragments can be used as templates.

3. Binary vector for expression of CRISPR/Cas9: Various CRISPR/Cas9 plasmids for use in rice are available from Addgene (https://www.addgene.org/). In the work described in this chapter, we utilize pZH_OsU6sgRNA_SpCas9-wt as the CRISPR/Cas9 expression vector [9].

4. Primers for amplification of sgRNA expression cassettes:

   sgRNA1 forward: 5′-**tctctaagcttggcgcgcc**ggtaccgagctcggatccac-3′

   sgRNA1 reverse: 5′-***ctatagtgagtcgtat***taaagcttatcgataccgtcga-3′

   sgRNA2 forward: 5′-***atacgactcactata***gggggtaccgagctcggatccac-3′

   sgRNA2 reverse: 5′-**ttacgaatttca**ggcgcgccaagcttatcgatacc-3′

   Extension sequences for In-Fusion cloning are indicated by bold. T7 promoter sequences (bold italic) are used as an extension to combine sgRNA1 and sgRNA2; thus, T7 promoter primers can be used for confirming the insertion of sgRNA cassettes into the binary vector.

5. Primers for amplification of the GT donor:

   GT donor forward: 5′-**caaacactgatagtttaaac**NNNNNNNNNNNNNNNNNNNNN -3′

   GT donor reverse: 5′-**ttcccgccttcagtttaaac**NNNNNNNNNNNNNNNNNNNN -3′

   "N" indicates target-specific sequences. Extension sequences for In-Fusion cloning are indicated by bold.

6. Primers for sequencing:

   sgRNA insertion check forward: 5′- taatacgactcactatagg-3′ (T7 promoter primer-F)

   sgRNA insertion check reverse: 5′- cctatagtgagtcgtatta-3′ (T7 promoter primer-R)

   GT donor insertion check forward: 5′- tgtcgcgtaacttaggacttgtg –3′

   GT donor insertion check reverse: 5′- cgtattggagcttgagcttgg – 3′

7. Restriction enzymes: *Asc*I and *Pme*I (New England Biolabs).

8. KOD One PCR Master Mix (TOYOBO).

9. QIAquick PCR Purification Kit (Qiagen).

10. Agarose gel electrophoresis equipment.

11. 1.0% agarose gel.

12. 1× Tris-acetate-EDTA (TAE) buffer.
13. In-Fusion HD Cloning Kit (Takara).
14. *Escherichia coli* DH5α competent cells.
15. QIAprep Spin Miniprep Kit (Qiagen).
16. Sanger sequencing kit.

2.2 Agrobacterium-Mediated Transformation

See Subheading 2.2 in Chap. 9.

2.3 Screening of Calli or Regenerated Plants Carrying the GT-Modified Target Locus

1. Agencourt Chloropure (Beckman Coulter).
2. Homogenizer, e.g., Multi-beads shocker (Yasui Kikai).
3. KOD One PCR Master Mix (TOYOBO).
4. Primers for GT confirmation: Prepare primer sets that amplify the fragment including both the entire 5′ or 3′ homology arm and the modification region. One primer should be located outside the GT donor template. For the inside primer, there are two alternatives. The first approach is to design the primer to extend beyond the modified region, ensuring the detection of modifications, such as large insertion or deletion (*see* **Note 2**). The second approach is to design the primer to anneal to the modification region itself, such as the inserted fragment (in this chapter, FLAG-tag), to detect modifications within the GT donor template (*see* Fig. 2). In the latter strategy, fragments can be amplified only when GT has occurred successfully. Also, prepare appropriate sequencing primers to verify the modification.
5. Agarose gel electrophoresis equipment or microchip electrophoresis device.
6. 1.0% agarose gel.
7. 1× TAE buffer.
8. Sanger sequencing kit.

3 Methods

3.1 All-in-One GT Vector Construction

We designed a two-step strategy for the construction of an all-in-one GT vector by first inserting two sgRNA expression cassettes, followed by inserting a GT donor template (*see* Fig. 3).

1. Digest >1 μg of pZH_OsU6sgRNA_SpCas9-wt with *Asc*I for cloning two sgRNA expression cassettes, following the manufacturer's protocol.
2. Purify the linearized vector using the QIAquick PCR Purification Kit. Elute with 50 μL of nuclease-free water.

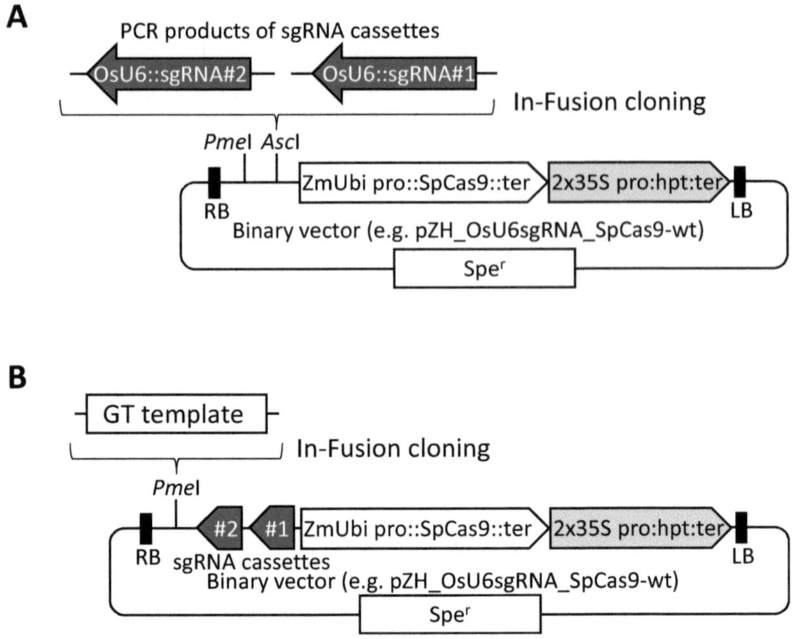

Fig. 3 Strategy for GT vector construction. (**a**) Schematic representation of cloning the sgRNA expression cassettes into the binary vector containing SpCas9 and HPT expression cassettes (e.g., pZH_OsU6sgRNA_SpCas9-wt). (**b**) Schematic representation of cloning the GT donor template into a binary vector obtained as a result of (**a**)

3. Confirm the complete digestion and quantify the concentration by agarose gel electrophoresis.
4. Amplify each sgRNA expression cassette using PCR. Assemble the PCR reaction as follows:

2× KOD One PCR Master Mix	25 μL
sgRNA1 or 2 forward primer (10 μM)	1.5 μL
sgRNA1 or 2 reverse primer (10 μM)	1.5 μL
sgRNA1 or 2 expression cassette template	~50 ng
Nuclease-free water	Up to 50 μL
Total	50 μL

5. Run PCR reaction as follows:

   2 min at 98 °C

   35 cycles of:

   10 s at 98 °C

   5 s at 60 °C

   5 s at 68 °C

   Hold at 12 °C

6. Purify the PCR products using the QIAquick PCR Purification Kit. Elute with 50 μL of nuclease-free water.
7. Confirm the amplification and quantify the concentration by agarose gel electrophoresis.
8. Insert the purified PCR products of sgRNA expression cassettes into the linearized vector from **step 2** by In-Fusion HD Cloning Kit according to the manufacturer's protocol (*see* **Note 3**).
9. Transform 1 μL of In-Fusion cloning reaction from **step 8** to *E. coli* DH5α competent cells according to the manufacturer's protocol.
10. Isolate plasmid DNA from individual colonies using the QIAprep Spin Miniprep Kit.
11. Confirm the insertion of two sgRNA expression cassettes and select positive clones, in which two sgRNA expression cassettes are inserted as expected, by sequencing using a Sanger sequencing kit and primers for sgRNA insertion check, following the manufacturer's protocol.
12. Digest >1 μg of pZH_OsU6sgRNA_SpCas9-wt carrying sgRNA expression cassettes with *Pme*I for cloning the GT donor template, following the manufacturer's protocol.
13. Purify the linearized vector using the QIAquick PCR Purification Kit. Elute with 50 μL of nuclease-free water.
14. Confirm the complete digestion and quantify the concentration by agarose gel electrophoresis.
15. (Optional; *see* **Note 4**) Add 20 bp overhang sequences to the GT donor template for In-Fusion cloning. Assemble the PCR reaction as follows:

2× KOD One PCR Master Mix	25 μL
GT donor forward primer (10 μM)	1.5 μL
GT donor reverse primer (10 μM)	1.5 μL
GT donor template	~50 ng
Nuclease-free water	Up to 50 μL
Total	50 μL

16. (Optional; *see* **Note 4**) Run PCR reaction as follows:

    2 min at 98 °C

    35 cycles of:

    10 s at 98 °C

    5 s at 60 °C

    5 s/kb (1–10 kb) at 68 °C

    Hold at 12 °C

17. (Optional; *see* **Note 4**) Purify the PCR products using the QIAquick PCR Purification Kit. Elute with 50 μL of nuclease-free water.
18. Insert the GT donor template with 20 bp overhang sequences into the linearized vector from **step 13** by In-Fusion HD Cloning Kit, following the manufacturer's protocol.
19. Transform 1 μL of In-Fusion cloning reaction from **step 18** to *E. coli* DH5α competent cells according to the manufacturer's protocol.
20. Isolate plasmid DNA from individual colonies using the QIAprep Spin Miniprep Kit.
21. Confirm the insertion of the GT donor template and select positive clones, in which the GT donor template is inserted as expected, by sequencing using a Sanger sequencing kit and primers for GT donor insertion check, following the manufacturer's protocol.

3.2 Agrobacterium-Mediated Transformation

Transform rice calli using the *Agrobacterium*-mediated transformation method (*see* Subheading 3.2 in Chap. 9).

3.3 Screening of Calli Carrying the Modified Target Locus via GT

1. Collect small pieces of hygromycin-resistant calli from more than 384 independent callus lines into collection tubes after 4 weeks of selection (*see* **Note 5**).
2. Grind callus with metal corns using homogenizer, e.g., Multibeads shocker.
3. Extract genomic DNA from these hygromycin-resistant calli using Agencourt Chloropure, following the manufacturer's protocol.
4. Amplify the GT target locus to identify the calli carrying the modified target locus via GT (insertion of a FLAG-tag sequence in this chapter). Assemble the PCR reaction as follows:

2× KOD One PCR Master Mix	5 μL
GT donor forward primer (10 μM)	0.3 μL
GT donor reverse primer (10 μM)	0.3 μL
Genomic DNA	~40 ng
Nuclease-free water	Up to 10 μL
Total	10 μL

5. Run PCR reaction as follows:

   2 min at 98 °C

   35 cycles of:

   10 s at 98 °C

   5 s at 60 °C

   5 s/kb (1–10 kb) at 68 °C

   Hold at 12 °C

6. Confirm the GT events by agarose gel electrophoresis or microchip electrophoresis (*see* **Note 6**).
7. Regenerate the PCR-positive calli (*see* Chap. 9).

3.4 Screening of Regenerated Plants Carrying the GT-Modified Target Locus

1. Individually transfer each regenerated plantlet from the calli selected in Subheading 3.3 with well-developed root systems from a hormone-free medium to the 15 mL tubes filled with water (*see* **Note 4**).
2. Extract DNA from small pieces of leaves of regenerated rice plants using Agencourt Chloropure, following the manufacturer's protocol.
3. Repeat **steps 4–6** in Subheading 3.3.
4. Purify the PCR products from the positive sample confirmed in **step 3**, using the QIAquick PCR Purification Kit.
5. Confirm the desired modification by GT by sequencing using a Sanger sequencing kit.
6. Transplant the regenerated plants in which the desired modification was verified in **step 5** to soil and grow in a greenhouse to obtain T_1 progenies.

4 Notes

1. If substitution mutations are introduced into the target gene via GT, additional mutations would be needed to avoid cleavage with CRISPR/Cas9 in the GT donor template.
2. If substitution mutations are introduced into the target gene via GT, sequencing of PCR products of the target locus is required to identify calli carrying the GT-modified target locus.
3. To avoid transcriptional interference, sgRNA expression cassettes should be located in a head-to-head orientation to the SpCas9 expression cassettes on the GT vector.
4. These 20 bp overhang sequences can be added to the GT donor template by gene synthesis or overlap extension PCR. Then, skip **steps 15–17** in Subheading 3.1.
5. The proportion of GT calli per hygromycin-resistant calli is approximately 0.2% (*see* Table 1) [5].

Table 1
Examples of GT experiments to introduce a 3 × Flag-tag into the N-terminal region of an endogenous rice gene

Ex	No. of hygromycin-resistant callus lines analyzed	No. of callus lines with Flag-tag knock-in allele	%	No. of lines that detected Flag-tag knock-in allele in regenerated plants	%
1	384	4	1.04	1	0.26
2	384	2	0.52	1	0.26

Table 2
Ratio of plants with Flag-tag knock-in allele to regenerated plants from GT-positive calli

Line no.	No. of plants analyzed	No. of plants with Flag-tag knock-in allele	%
1	24	24	100
2	24	9	37.5

6. The GT event is thought to occur in a part of the PCR-positive callus; thus, not all plants regenerated from PCR-positive callus harbor the GT-modified target gene. At least 24 plants should be subjected to PCR screening to isolate GT-positive plants (*see* Table 2).

References

1. Puchta H, Fauser F (2013) Gene targeting in plants: 25 years later. Int J Dev Biol 57:629–637
2. Chen K, Wang Y, Zhang R et al (2019) CRISPR/Cas genome editing and precision plant breeding in agriculture. Annu Rev Plant Biol 70:667–697
3. Terada R, Urawa H, Inagaki Y et al (2002) Efficient gene targeting by homologous recombination in rice. Nat Biotechnol 20:1030–1034
4. Shimatani Z, Nishizawa-Yokoi A, Endo M et al (2014) Positive-negative-selection-mediated gene targeting in rice. Front Plant Sci 5:748
5. Nishizawa-Yokoi A, Mikami M, Toki S (2020) A universal system of CRISPR/Cas9-mediated gene targeting using all-in-one vector in plants. Front Genome Ed 2:14
6. Li J, Kong D, Ke Y et al (2024) Application of multiple sgRNAs boosts efficiency of CRISPR/Cas9-mediated gene targeting in Arabidopsis. BMC Biol 22:6
7. Endo M, Nishizawa-Yokoi A, Toki S (2016) Targeted mutagenesis in rice using TALENs and the CRISPR/Cas9 system. In: Murata M (ed) Chromosome and genomic engineering in plants: methods and protocols. Springer, New York, pp 123–135
8. Endo M, Nishizawa-Yokoi A, Toki S (2018) Rice genome editing. In: Sasaki T, Ashikari M (eds) Rice genomics, genetics and breeding. Springer, Singapore, pp 523–539
9. Endo M, Mikami M, Endo A et al (2019) Genome editing in plants by engineered CRISPR-Cas9 recognizing NG PAM. Nat Plants 5:14–17

Chapter 11

Precise Base Substitution Using CRISPR/Cas-Mediated Base Editor in Rice

Masaki Endo, Katsuya Negishi, and Seiichi Toki

Abstract

Base editors, CRISPR/Cas-based precise genome editing tools, enable base conversion at a target site without inducing DNA double-strand breaks. The genome editing targetable range is restricted by the requirement for protospacer adjacent motif (PAM) sequence. Cas9 derived from *Streptococcus pyogenes* (SpCas9)—most widely used for genome editing in many organisms—requires an NGG sequence adjacent to the target site as a PAM. Then, engineered and natural Cas variants with altered PAM recognition are used for base editor to expand the flexibility of base substitution position. In this chapter, we describe a protocol for base editing based on SpCas9-NG, which is a rationally engineered SpCas9 variant that can recognize relaxed NG PAMs.

Key words Cytosine base editors, Adenine base editors, Base substitution, *Agrobacterium*-mediated transformation, CRISPR/Cas

1 Introduction

As DNA sequencing technology has evolved and comparative genomic analysis has progressed, it has become revealed that differences in agronomically important traits are often derived from single nucleotide polymorphisms. Therefore, precise genome editing technologies enabling to substitute specific nucleotides have been required for accelerating the breeding process. Currently, two base editors, cytosine base editors (CBEs) for C:G to T:A transition and adenine base editors (ABEs) for A:T to G:C transition, are widely used in plants. Furthermore, base editors that combine both cytosine and adenine base editing functions were developed [1–3]. The above base editors utilize deaminase for substitution, so that substitution pattern is nucleotide transition. For transversions, glycosylase-based base editors are developed; C-to-G base editors (CGBEs) for C:G to G:C transversion, A-to-Y base editor (AYBE)

Taiji Kawakatsu (ed.), *Rice: Methods and Protocols*, Methods in Molecular Biology, vol. 2869,
https://doi.org/10.1007/978-1-0716-4204-7_11,

for A:T to T:A or C:G transversion, and A-to-K base editor (AKBE) for A:T to G:C or T:A transversion have been developed [4–7].

In this chapter, we describe base substitution methods using CBE and ABE. CBE is composed of a SpCas9 nickase (nSpCas9) fused with a catalytic domain of cytidine deaminase and uracil glycosylase inhibitor (UGI) and single guide RNA (sgRNA). CBE binds to the target sequence in the genomic DNA in a sgRNA-programmed manner. CBE converts the C to U (uracil) at a target site. U is recognized as thymine (T) during DNA replication, resulting in a C:G to T:A transition (*see* Fig. 1a). UGI inhibits the excision of the U by uracil DNA glycosylase (UDG), which leads to reversion to the C:G starting state. The first CBE was designed in 2016 [8], and since then, various improvements have been made to increase efficiency and specificity [9]. The protocol described in this chapter is the way of C to T substitution using SpCas9-NG [10] (*see* **Note 1**) nickase (nSpCas9-NG) fused with the *Petromyzon marinus* cytidine deaminase (PmCDA1) and UGI [11, 12].

ABE is composed of a nSpCas9 fused with an engineered deoxyadenosine deaminase (TadA) and sgRNA. ABE converts adenine (A) to inosine (I) resulting in an A:T to G:C transition (*see* Fig. 1b). The initial ABE was developed in 2017 [13], and numerous optimizations and modifications were performed to enhance their editing efficiency [9]. The protocol described in this chapter is the way of A to G substitution using ABE8e [14, 15] with nSpCas9-NG. The base editing windows of CBE based on SpCas9 and PmCDA1 are around −16 to −20 from PAM, and that of ABE8e using SpCas9 are located at −13 to −17 from PAM (*see* Fig. 2 and **Note 2**). Therefore, it is necessary to first ascertain the positional relationship between the PAM sequence and the Cs or As. In some cases, multiple Cs or As in the edit window are replaced simultaneously, while in other cases, only single base at specific position is replaced.

2 Materials

2.1 Vector Construction

1. Oligonucleotides for sgRNA: Select a 20-nt sequence upstream PAM sequence as a target sequence. SpCas9-NG recognizes 5′-NG-3′ as a PAM sequence. Prepare two oligonucleotides as follows (*see* Fig. 3a and **Notes 3** and **4**):

   sgRNA forward oligo: 5′-GTTGNNNNNNNNNNNNNNNNNNNN-3′ (*N* = 20-nt target sequence)

   sgRNA reverse oligo: 5′-AAACNNNNNNNNNNNNNNNNNNNN-3′ (*N* = 20-nt reverse and complementary target sequence)

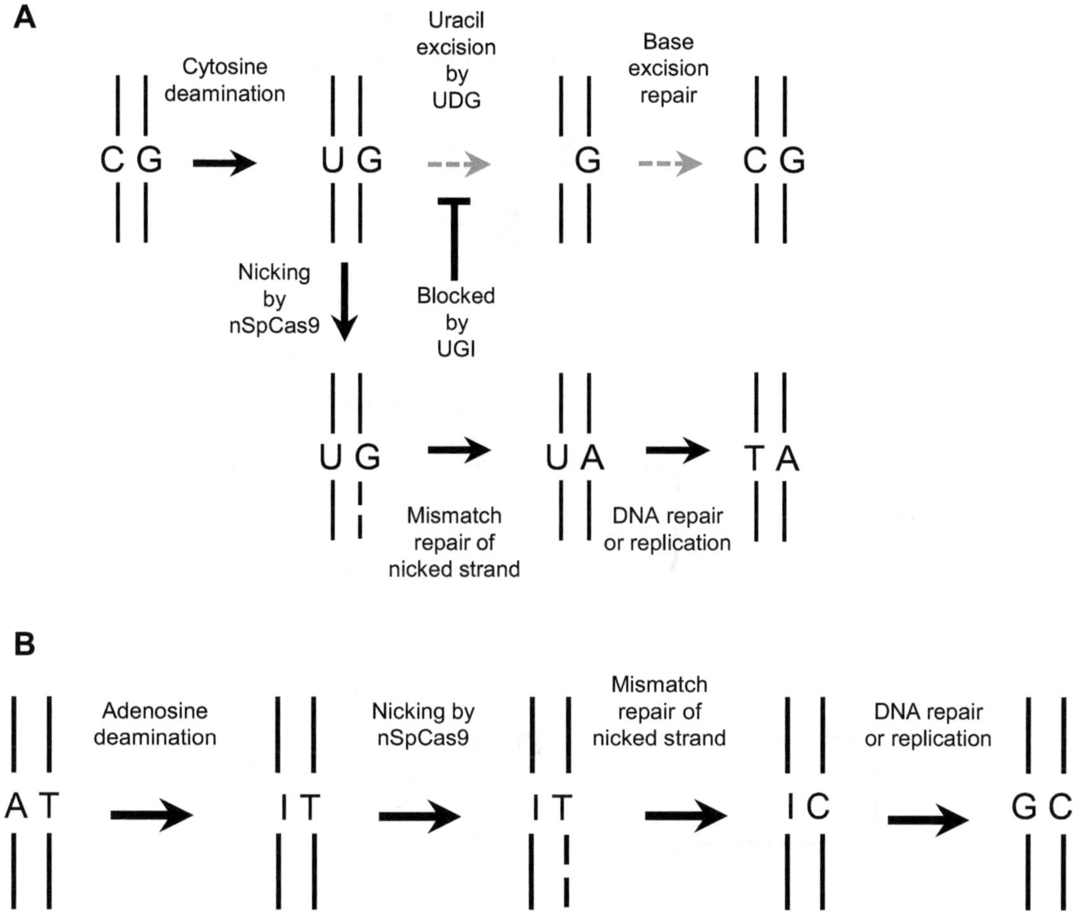

Fig. 1 Schematic representation of CBE- and ABE-mediated nucleotide substitution. (**a**) CRISPR/nCas9-mediated cytosine base editing. The cytidine deaminase catalyzes the deamination of cytosine (C) and makes the base change from C to uracil (U). When U:G mismatch occurs, U is removed by uracil DNA glycosylase (UDG) in nature. However, with CBE, UDG activity is inhibited by uracil glycosylase inhibitor (UGI), and the U:G mismatch is processed by mismatch repair. The damaged strand is preferentially repaired so that G at the nicked strand is replaced to A resulting in a C·G to T·A transition. (**b**) CRISPR/nCas9-mediated adenine base editing. The adenosine deaminase catalyzes the deamination of adenine (A) and makes the base change from A to inosine (I). I:T mismatch is processed by mismatch repair, which preferentially repairs the nicked strand, and a mismatched T is replaced to C. I is recognized G during DNA replication, resulting in a A·T to G·C transition

	-20	-19	-18	-17	-16	-15	-14	-13	-12	-11	-10	-9	-8	-7	-6	-5	-4	-3	-2	-1	PAM		
5'-	C	C	C	C	C	N	N	N	N	N	N	N	N	N	N	N	N	N	N	N	**N**	**G**	-3'
5'-	N	N	N	A	A	A	A	A	N	N	N	N	N	N	N	N	N	N	N	N	**N**	**G**	-3'

Editing window of CBE vector, pPZP_ZmUbi_nSpCas9-NG_PmCDA_UGI

Editing window of ABE vector, pPZP_ZmUbi_ABE8e_nSpCas9-NG

Fig. 2 Editing windows of CBE and ABE. Base editors bind to the target sites in a sgRNA-programmed manner. CBE vector used in this protocol converts Cs located between −16 and −20, relative to PAM, to Ts. ABE vector used in this protocol converts As located between −13 and −17, relative to PAM, to Gs

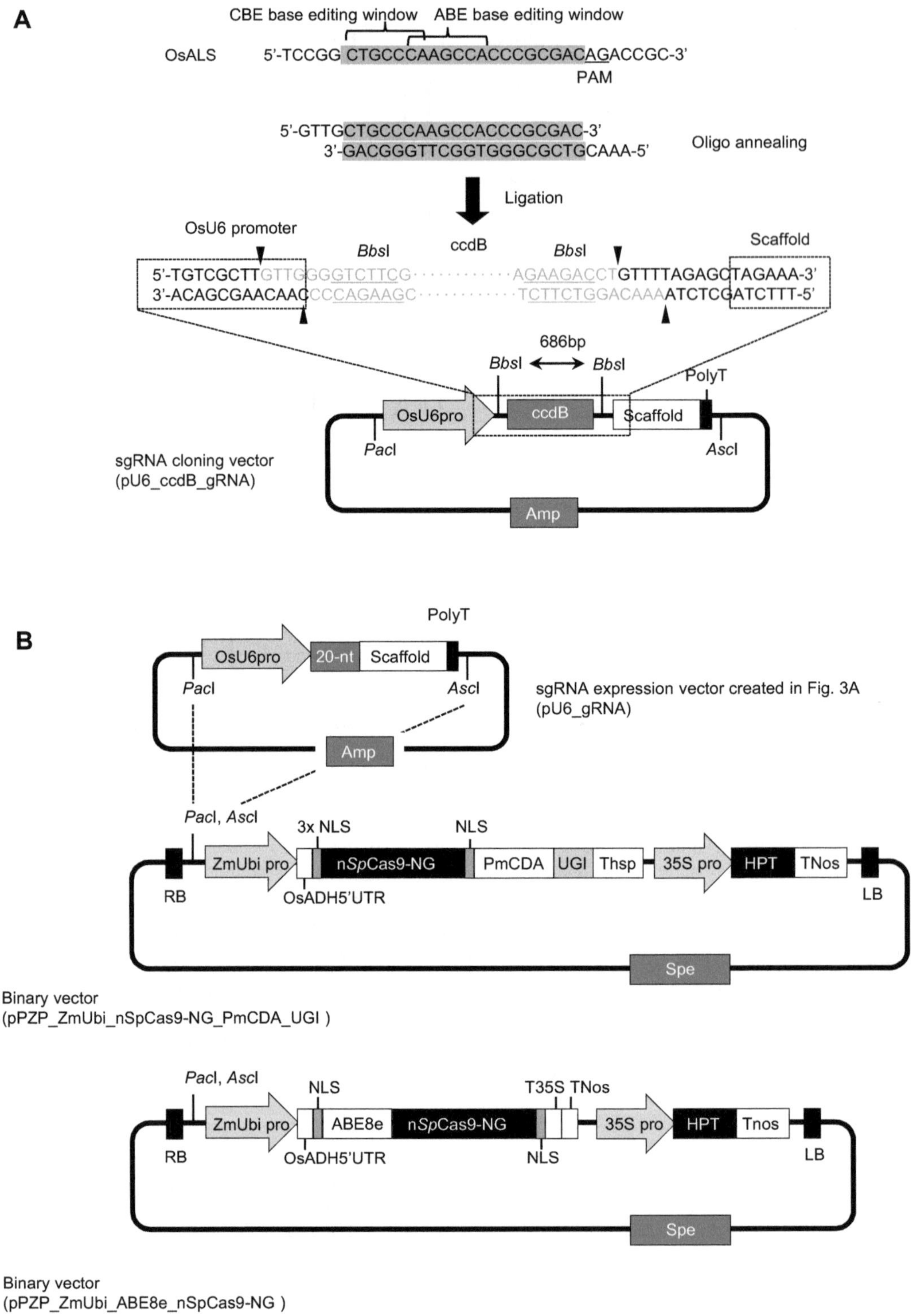

Fig. 3 Schematic representation of vector construction. (**a**) First, for a sgRNA expression cassette, two oligonucleotides are prepared. For example, if you want to induce a C to T substitution in the region of the CBE base editing window shown at the top of this diagram, synthesise the following two oligos, sgRNA forward oligo: 5′-GTTGCTGCCCAAGCCACCCGCGAC-3′ and sgRNA reverse oligo: 5′-AAACGTCGCGGGTGGCTTGGGCAG-3′. The underlined "AG" is the PAM in this case. The gray highlighted

2. sgRNA cloning vector: pU6_ccdB_gRNA containing the OsU6 promoter and the sgRNA scaffold of SpCas9 (*see* Fig. 3a).
3. Binary vector for CBE: pPZP_ZmUbi_nSpCas9-NG_PmCDA_UGI containing expression cassette of SpCas9-NG nickase fused with the catalytic domain of the PmCDA1 and UGI and hygromycin phosphotransferase (HPT) expression cassette (*see* Fig. 3b).
4. Binary vector for ABE: pPZP_ZmUbi_ABE8e_nSpCas9-NG containing nSpCas9-NG fused with the ABE8e and HPT expression cassette (*see* Fig. 3b).
5. Restriction enzymes: AscI, PacI, BbsI (New England Biolabs).
6. T4 DNA Ligation Kit (Takara): This kit includes 10× ligation buffer and T4 DNA ligase.
7. *E. coli* competent cell: *ccd*B survival strains such as One Shot *ccd*B Survival 2 T1R Chemically Competent Cells (Invitrogen) are needed to maintain pU6_ccdB_gRNA. For the other step, *E. coli* competent cell such as DH5α strain is used.
8. LB broth, Miller (Nacalai): Prepare both LB solid and liquid media containing 100 μg/mL ampicillin or spectinomycin. Autoclave at 121 °C for 20 min and store at 4 °C.
9. KOD One PCR Master Mix (TOYOBO).
10. QIAquick Gel Extraction Kit (Qiagen).
11. QIAprep Spin Miniprep Kit (Qiagen).
12. BigDye Terminator v3.1 cycle sequencing kit (Applied Biosystems).
13. Primers for sequencing:

    M13 forward: 5′-GTAAAACGACGGCCAGT-3′

    M13 reverse: 5′-CAGGAAACAGCTATGAC-3′
14. 1× TAE: 40 mM Tris–HCl, 20 mM acetic acid, 1 mM EDTA.
15. 1% agarose gel in 1× TAE.

Fig. 3 (continued) 20 nt is the target sequences to be placed on the sgRNA. Next, annealed oligonucleotides are ligated into the BbsI sites of the pU6_ccdB_gRNA for making a sgRNA expression cassette. Nucleotides in light gray indicate sequences removed by BbsI digestion. (**b**) The resultant sgRNA expression cassette is ligated into the binary vectors harboring SpCas9-NG nickase, deaminase fusion protein, and selection marker by using AscI and PacI sites. Binary vector pPZP_ZmUbi_nSpCas9-NG_PmCDA_UGI is for CBE and pPZP_ZmUbi_ABE8e_nSpCas9-NG is for ABE. OsU6 pro, ZmUbi pro, and 35S pro, rice U6–2 promoter; maize ubiquitin 1 promoter, and cauliflower mosaic virus 35S RNA promoter, respectively. OsADH 5′-UTR; the 5′ of an untranslated region of rice alcohol dehydrogenase 2 gene as a translational enhancer. NLS; the nuclear localization signal derived from Simian virus 40. HPT; hygromycin phosphotransferase. Thsp, Tnos, and T35S; the terminator region of rice heat shock protein 17.3, nopaline synthase, and cauliflower mosaic virus 35S RNA, respectively. LB and RB; left and right border sequences, respectively. Amp and Spe; ampicillin and spectinomycin resistance genes, respectively

2.2 *Agrobacterium*-Mediated Transformation

See Subheading 2.2 in Chap. 9.

2.3 DNA Extraction and Mutation Analysis

1. Agencourt Chloropure (Beckman Coulter).
2. KOD One PCR Master Mix (TOYOBO).
3. BigDye Terminator v3.1 cycle sequencing kit (Applied Biosystems).
4. Target-specific primers: Design PCR primers to amplify 400–500 bp around the target site of CBE or ABE.
5. QIAquick PCR Purification Kit (Qiagen).

3 Methods

3.1 Vector Construction

The outline of the vector construction is shown in Fig. 3.

1. Digest pU6_ccdB_gRNA by BbsI. Assemble the restriction enzyme reaction as follows:

10× NEBuffer 2.1	5 μL
pU6_ccdB_gRNA	1–44 μL (2 μg)
BbsI (10 units/μL)	1 μL
H_2O	Up to 50 μL
Total	50 μL

2. Incubate the restriction enzyme reaction from **step 1** at 37 °C for 2–16 h.
3. Perform agarose electrophoresis. The linearized DNA fragments should separate into 3.2 kb and 0.7 kb.
4. Purify the 3.2 kb fragment using the Gel Extraction Kit following the manufacturer's protocols (*see* **Note 5**). Adjust the concentration to 10 ng/μL with sterile distilled water.
5. Mix 2 μL of 50 μM each strand of oligonucleotide for gRNA and 46 μL sterile distilled water. Heat the mixture at 95 °C for 5 min to denature the oligonucleotides and cool it down to the room temperature slowly and incubate at room temperature for 15–30 min to anneal the oligonucleotides.

6. Assemble the ligation reaction as follows:

10× ligation buffer	1 μL
Linearized pU6_ccdB_gRNA from **step 4**	2 μL
T4 DNA ligase	1 μL
H_2O	6 μL
Total	10 μL

7. Incubate the ligation reaction from **step 6** at room temperature for 1 h.
8. Transform ligation products into *E. coli* competent cells following the manufacturer's protocols. Spread transformed *E. coli* to the LB solid medium containing 100 μg/mL ampicillin for selection. Incubate at 37 °C overnight, and then store at 4 °C.
9. Inoculate a single colony of ampicillin-resistant *E. coli* from **step 8** into 2 mL of LB liquid medium containing 100 μg/mL ampicillin and incubate at 37 °C with shaking overnight.
10. Isolate the plasmid from the cultured *E. coli* by using QIAprep Spin Miniprep Kit following the manufacturer's protocol.
11. Confirm the successful cloning of 20-nt target sequence by sequencing using BigDye Terminator v3.1 cycle sequencing kit and M13 forward or reverse primers following the manufacturer's protocol. At this step, construction of the sgRNA expression vector, pU6_gRNA (*see* Fig. 3b), is completed.
12. Digest pU6_gRNA and pPZP_ZmUbi_nSpCas9-NG_PmCDA_UGI for CBE or pPZP_ZmUbi_ABE8e_nSpCas9-NG for ABE (*see* Fig. 3b). Assemble the restriction enzyme reaction for each plasmid as follows:

10× CutSmart buffer	5 μL
Plasmid	1–43 μL (2 μg)
AscI	1 μL
PacI	1 μL
H_2O	Up to 50 μL
Total	50 μL

13. Incubate the restriction enzyme reaction from **step 12** at 37 °C for 2–6 h.
14. Perform agarose electrophoresis. Sizes of the linearized DNA fragments should be 0.5 kb (pU6_gRNA), 17.2 kb (pPZP_ZmUbi_nSpCas9-NG_PmCDA_UGI), and 16.4 kb (pPZP_ZmUbi_ABE8e_nSpCas9-NG).

15. Purify the linearized fragment using the Gel Extraction Kit following the manufacturer's protocols.
16. Assemble the ligation reaction as follows:

10× ligation buffer	1 μL
Linearized pPZP from **step 15**	1–4 μL (60 ng)
Linearized pU6_gRNA from **step 15**	1–4 μL (5 ng)
T4 DNA ligase	1 μL
H_2O	Up to 10 μL
Total	10 μL

17. Incubate the ligation reaction from **step 16** at room temperature for 1 h.
18. Transform ligation products into *E. coli* competent cells following the manufacturer's protocols. Spread transformed *E. coli* to the LB solid medium containing 100 μg/mL spectinomycin for selection. Incubate at 37 °C overnight, and then store at 4 °C.
19. Inoculate a single colony of spectinomycin-resistant *E. coli* from **step 18** into 2 mL of LB liquid medium containing 100 μg/mL spectinomycin and incubate at 37 °C with shaking overnight.
20. Isolate the plasmid from the cultured *E. coli* by using QIAprep Spin Miniprep Kit following the manufacturer's protocol.
21. Digest the plasmid to confirm the insertion of the sgRNA expression cassette. Assemble the restriction enzyme reaction for each plasmid as follows:

10× CutSmart buffer	1 μL
Plasmid	1–8 μL (200 ng)
AscI	0.5 μL
PacI	0.5 μL
H_2O	Up to 10 μL
Total	10 μL

22. Incubate the restriction enzyme reaction from **step 21** at 37 °C for 1 h.
23. Perform agarose electrophoresis. If the sgRNA expression cassette is inserted, the fragment should separate into 0.5 kb (the sgRNA expression cassette) and 17 kb (the pPZP binary vector backbone).

3.2 Rice Transformation

Transform rice calli using the *Agrobacterium*-mediated transformation method (*see* Subheading 3.2 in Chap. 9).

3.3 Mutation Analysis

1. Harvest the small pieces of hygromycin-resistant calli after 5–6 weeks of selection and extract genomic DNA from them using Agencourt Chloropure, following the manufacturer's protocol.
2. Amplify the target locus using genomic DNA extracted at **step 1** and target specific primers with DNA polymerase following the manufacturer's protocols. Prepare the PCR reaction mix as follows:

Genomic DNA	5 μL
10 μM target-specific forward primer	1.5 μL
10 μM target-specific reverse primer	1.5 μL
KOD one PCR master mix	25 μL
H_2O	17 μL
Total	50 μL

3. Run the thermal cycler reaction as follows:

   2 min at 98 °C

   35 cycles of:

   10 s at 98 °C

   5 s/kb (1–10 kb) at 68 °C

   7 min at 68 °C

   Hold at 12 °C
4. Load 2 μL of PCR products on the gel to confirm the amplification of single PCR product.
5. Purify the PCR products using QIAquick PCR Purification Kit following the manufacturer's protocol.
6. Confirm the sequences of purified PCR product by sequencing using Sanger sequencing kit and target specific primers used in **step 2**, according to the manufacturer's protocol (*see* **Note 6**).

4 Notes

1. For CRISPR editing tools, a highly conserved PAM sequence (e.g., NGG PAM for the widely used SpCas9) would greatly restrict targetable regions in the genome. Evolving Cas variants with relaxed PAM recognition is a major strategy for expanding the scope of genome editing. A rationally engineered SpCas9

variant named SpCas9-NG with R1335V/L1111R/D1135V/G1218R/E1219F/A1322R/T1337R (VRVRFRR) exhibited robust recognition of NGN PAM in the genomes of eukaryotic cells [10]. A further D10A amino acid substitution results in the SpCas9-NG nickase.

2. The editing window for base substitution varies slightly depending on the type of deaminase, how it is fused to nSpCas9-NG, and the length of the linker.
3. Select target sequences and design oligonucleotides for sgRNA vector construction (*see* Fig. 3a). Base substitutions with CBE or ABE can be introduced to C or A, respectively, at a certain distance from the PAM sequence (*see* Fig. 2). Therefore, the presence of a PAM sequence (NG in the case of SpCas9-NG) at an appropriate distance away from the C or A for which base substitution is desired is a prerequisite for the experiment.
4. Nishimasu et al. [10] showed that relative to wild-type SpCas9, SpCas9-NG slowly but more efficiently cleaved the TGA, TGT, and TGC in vitro. For the TGG PAM target site, in vitro cleavage occurs more rapidly with SpCas9-WT than with SpCas9-NG target site [10].
5. Because *ccd*B encodes a toxic protein that acts as a DNA gyrase toxin, *ccd*B-resistant *E. coli* such as One Shot *ccd*B Survival 2 T1R Chemically Competent Cells must be used to amplify pU6_ccdB_gRNA. The presence of the *ccd*B gene between the two BbsI sites makes it easier to confirm by electrophoresis that the restriction enzyme treatment is complete. Also, if plasmid with incomplete BbsI digestion is used for cloning the 20-nt target sequence, the remaining *ccd*B sequence kills the non--*ccd*B-resistant *E. coli* making it easier to select clones in which *ccd*B has replaced the target sequence.
6. Mutations in callus often occur in a chimeric fashion, so in most cases, base substitutions are detected as double peaks in Sanger sequencing. The callus lines with a high mutation rate, as determined by the sequencing analysis, are selected and preferentially used to generate regenerated plants. Performing mutation analysis at the callus stage makes it more efficient to obtain individuals with the desired base substitution.

References

1. Sakata RC, Ishiguro S, Mori H et al (2020) Base editors for simultaneous introduction of C-to-T and A-to-G mutations. Nat Biotechnol 38:865–869
2. Grünewald J, Zhou R, Lareau CA et al (2020) A dual-deaminase CRISPR base editor enables concurrent adenine and cytosine editing. Nat Biotechnol 38:861–864
3. Zhang X, Zhu B, Chen L et al (2020) Dual base editor catalyzes both cytosine and adenine base conversions in human cells. Nat Biotechnol 38:856–860

4. Kurt I, Zhou R, Iyer S et al (2021) CRISPR C-to-G base editors for inducing targeted DNA transversions in human cells. Nat Biotechnol 39:41–46
5. Zhao D, Li J, Li S et al (2021) Glycosylase base editors enable C-to-A and C-to-G base changes. Nat Biotechnol 39:35–40
6. Li Y, Li S, Li C et al (2023) Engineering a plant A-to-K base editor with improved performance by fusion with a transactivation module. Plant Commun 4:100667
7. Tong H, Wang X, Liu Y et al (2023) Programmable A-to-Y base editing by fusing an adenine base editor with an N-methylpurine DNA glycosylase. Nat Biotechnol 41:1080–1084
8. Komor A, Kim YB, Packer MS et al (2016) Programmable editing of a target base in genomic DNA without double-stranded DNA cleavage. Nature 533:420–424
9. Li J, Zhang C, He Y et al (2022) Plant base editing and prime editing: the current status and future perspectives. J Integrative Plant Biol 65:444–467
10. Nishimasu H, Shi X, Ishiguro S et al (2018) Engineered CRISPR-Cas9 nuclease with expanded targeting space. Science 361:1259–1262
11. Nishida K, Arazoe T, Yachie N et al (2016) Targeted nucleotide editing using hybrid prokaryotic and vertebrate adaptive immune systems. Science 353:aaf8729
12. Shimatani Z, Kashojiya S, Takayama M et al (2017) Targeted base editing in rice and tomato using a CRISPR-Cas9 cytidine deaminase fusion. Nat Biotechnol 35:441–443
13. Gaudelli NM, Komor AC, Rees HA et al (2017) Programmable base editing of A*T to G*C in genomic DNA without DNA cleavage. Nature 551:464–471
14. Lipinaite A, Knott GJ, Palumbo CM et al (2020) DNA capture by a CRISPR-Cas9–guided adenine base editor. Science 369:566–571
15. Richter MF, Zhao KT, Eton E et al (2020) Phage-assisted evolution of an adenine base editor with improved Cas. Nat Biotechnol 38:883–891

Chapter 12

DeLTa-Seq: High-Throughput Targeted RNA-Seq of Rice Leaves Without RNA Purification

Makoto Kashima, Yasuyuki Nomura, and Atsushi J. Nagano

Abstract

DeLTa-seq is a high-throughput RNA-seq library preparation method that enables quantification of the expression of hundreds of arbitrarily selected genes without RNA purification. This method involves direct reverse transcription using rice leaf lysate and targeted RNA-seq library preparation. DeLTa-seq enables the precise quantification of gene expression with a small number of sequencing reads. This chapter provides detailed information on the design of gene-specific primers, sampling of rice leaves, preparation of lysates, direct-lysate reverse transcription, targeted RNA-seq library preparation, and bioinformatic analysis of DeLTa-seq data.

Key words Direct-lysate reverse transcription, Targeted RNA-seq, Gene expression analysis, Rice, Leaf, High-throughput RNA-seq

1 Introduction

In plant biology, including rice research, large-scale transcriptome analyses of numerous samples using RNA-seq have recently emerged [1–3]. Although the cost of sequencing has reduced markedly over the last decade [4], library preparation for RNA-seq is time-consuming and demanding, hindering its extensive use in routine applications. Although several high-throughput 3′ RNA-seq library preparation methods have been introduced [5–7], RNA extraction from rice samples is typically the most time-consuming and expensive step in RNA-seq, and eliminating the process could decrease the required labor and cost of the procedure significantly. Additionally, focusing on a limited number of genes of interest enables more accurate gene expression quantification with few sequencing reads owing to the trade-off between comprehensive gene quantification and the required amount of sequence data [8].

Taiji Kawakatsu (ed.), *Rice: Methods and Protocols*, Methods in Molecular Biology, vol. 2869,
https://doi.org/10.1007/978-1-0716-4204-7_12,

In this chapter, we present "direct-lysate targeted RNA-seq" (DeLTa-seq), a method that reduces the cost and labor required for RNA-seq of rice samples significantly while ensuring precise quantification. DeLTa-seq could enable the examination of biological phenomena with higher time resolution, space resolution, perturbation, and individual variability in rice under fluctuating field conditions.

2 Materials

2.1 Design of Gene-Specific Primers

Gene-specific primers for *O. sativa* cDNA sequences can be designed using Primer3 (version 0.4.0) [9] with the following parameters: concentration of monovalent cations = 40 mM, concentration of divalent cations = 6 mM, annealing oligo concentration = 25 nM, primer size min/opt/max = 20/23/25, primer Tm min/opt/max = 60/65.5/70, primer GC% min/opt/max = 30/50/55, max Tm difference = 5, product size ranges = 289–889, number to return = 5, PRIMER_RIGHT_INPUT = CAAGCAGAAGACGGCATACGAGAT , PRIMER_MAX_END_STABILITY. Candidate gene-specific primers are then concatenated as follows: TCGTCGGCAGCGTCAGATGTGTATAAGAGACAG + gene-specific primers. Candidate gene-specific primers for *O. sativa* are available at https://sites.google.com/view/delta-seq/.

2.2 Sampling of Rice Leaves

1. 2 mL microtube.
2. 4 mm zirconia beads.
3. Liquid nitrogen.

2.3 Preparation of Lysate

1. TissueLyser II.
2. TissueLyser adapter sets.
3. Direct-lysate reverse transcription (RT) buffer: 90 mM Tris–HCl, pH 7.6, 100 mM dithiothreitol.

2.4 Quality Check of the RNA in the Lysate (Optional; See Note 1)

1. PCR tube.
2. AMPure XP beads.
3. MagnaStand.
4. 80% ethanol (EtOH).
5. Nuclease-free water.
6. Quantus Fluorometer.
7. QuantiFluor RNA System.
8. 1% agarose gel or Agilent Bioanalyzer 2100 and RNA nano- or pico-kit.

2.5 Direct-Lysate Reverse Transcription

1. SuperScript IV Reverse Transcriptase.
2. RT primers: CAGAAGACGGCATACGAGATxxxxxxxxGT-GACTGGAGTTCAGACGTGTGCTCTTCC-GATCNNNNNNNNNNTTTTTTTTTTTTTTTTTTTTV. xxxxxxxx indicates index sequences for multiplex sequencing. NNNNNNNNNN indicates 10-base unique molecular identifier (UMI). The information of the RT primers is available at https://sites.google.com/view/lasy-seq/.
3. 25 mM dNTP.
4. Wizard SV Gel and PCR Clean-Up System.
5. Nuclease-free water.
6. Thermal cycler.

2.6 Targeted RNA-Seq Library Preparation

1. RNase T1.
2. RNase A.
3. AMPure XP beads.
4. Nuclease-free water.
5. KAPA HiFi HotStart ReadyMix.
6. 10 mM Tris–HCl, pH 7.6.
7. Pools of 10 μM gene-specific primers from Subheading 2.1.
8. EvaGreen dye, 20× in water.
9. Second index primers: AATGATACGGCGACCACCGA-GATCTACACxxxxxxxxTCGTCGGCAGCGTC. xxxxxxxx indicates the index sequences for multiplex sequencing. The information of the second index primers is available at https://sites.google.com/view/lasy-seq/.
10. PCR_P7 primer: CAAGCAGAAGACGGCATACGAGAT.
11. Thermal cycler.
12. Real-time thermal cycler.
13. Quantus Fluorometer.
14. QuantiFluor ONE dsDNA System.
15. Agilent Bioanalyzer 2100.
16. Agilent High-Sensitivity DNA kit.

2.7 Bioinformatic Analysis of DeLTa-Seq

1. fastp [9].
2. bowtie [10].
3. samtools [11].
4. UMI-tools [12].
5. RSEM [13].

3 Methods

All procedures are carried out at room temperature (15–25 °C) unless otherwise specified.

3.1 Sampling the Rice Leaves

1. Cut each *O. sativa* leaf with scissors and placed in a 2 mL microtube containing a few zirconia beads.
2. Snap-freeze in liquid nitrogen. Store at −80 °C until pulverization.

3.2 Lysate Preparation

1. Chill 2 mL TissueLyser II adapter for >30 min at −80 °C.
2. Set the sample tube in the pre-chilled adapter and pulverize the leaf for 30 s using TissueLyser II by increasing the speed from the lowest to the highest. Snap-freeze in liquid nitrogen.
3. Repeat **step 2** until the leaves are completely pulverized.
4. Add 400 μL of direct-lysate RT buffer, immediately followed by vortexing for a few minutes (*see* Fig. 1a). The lysate can be stored at −80 °C for at least 1 month (*see* **Note 2**).

3.3 Quality Check the RNA in the Lysate (Optional; See Note 1)

1. Transfer part of the lysate to a new PCR tube.
2. Add 2.5× volume of AMPure XP beads and vortex well.
3. Incubate at room temperature for 5 min.
4. Place the tube on a magnetic stand for 3 min.
5. Discard the supernatant.
6. Add 150 μL of 80% EtOH and discard the supernatant.

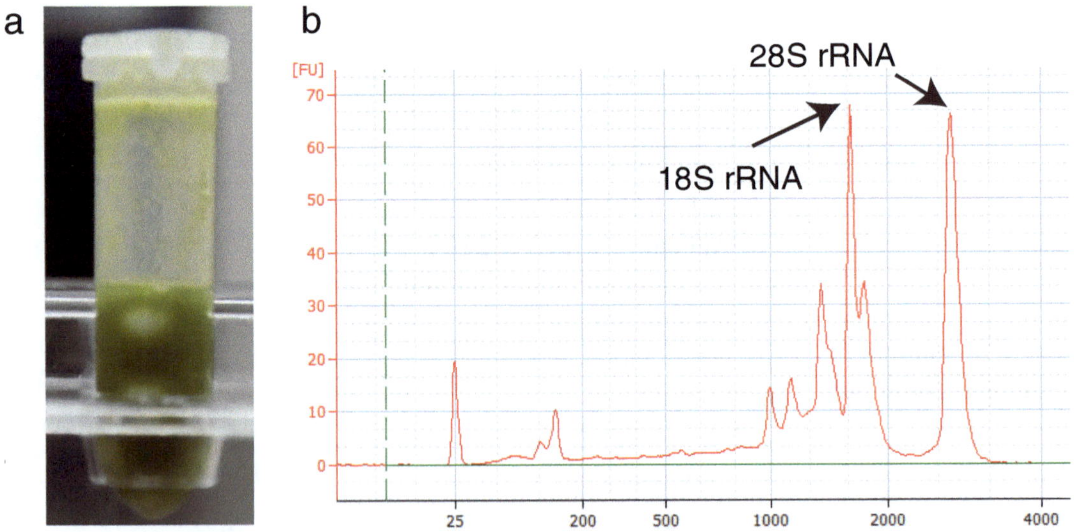

Fig. 1 Lysate of *Oryza sativa* leaf in direct-lysate RT buffer. (**a**) An image of the *O. sativa* leaf lysate. (**b**) A Bioanalyzer electropherogram of purified RNA from the lysate of *O. sativa* leaves

7. Repeat **step 6**.
8. Add an equal volume of nuclease-free water to a portion of the lysate and vortex well.
9. Incubate at room temperature for 1 min.
10. Place the tube on a magnetic stand for 3 min.
11. Collect the supernatant in a new PCR tube.
12. Measure the concentration of the purified RNA solution using Quantus Fluorometer and the QuantiFluor RNA System.
13. Electrophorese the RNA solution using 1% agarose gel or Agilent Bioanalyzer 2100 with an Agilent RNA nano- or pico-kit to verify the quality (*see* Fig. 1b).

3.4 Direct-Lysate Reverse Transcription

1. Prepare the reverse transcription (RT) reaction mix using RT primers with distinct index for each sample as follows:

2 μM RT primer	1 μL
25 mM dNTP	0.4 μL
5× SSIV buffer	4.0 μL
SuperScript IV reverse transcriptase	0.1 μL
Lysate	1 μL
Nuclease-free water	13.5 μL
Total	20 μL

2. Run the RT reaction as follows:

   10 min at 65 °C

   15 min at 80 °C

   hold at 4 °C

3. Pool all RT reaction mix and purify RT products using the Wizard SV Gel and PCR Clean-Up System according to the manufacturer's instructions, except for using 80% EtOH instead of the membrane wash solution. Elute the purified cDNA with 23 μL nuclease-free water.

The library structure is illustrated in Fig. 2.

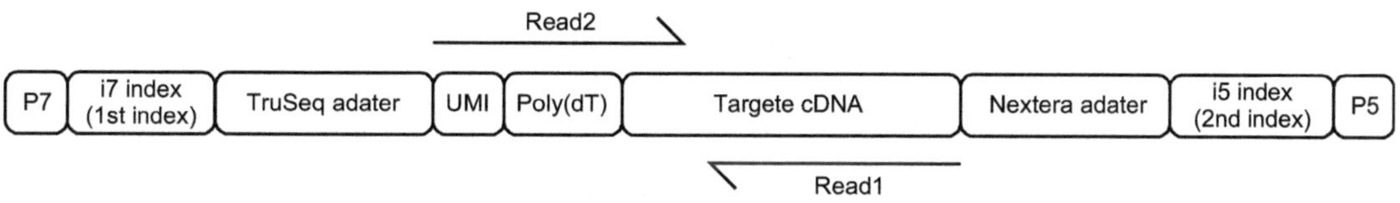

Fig. 2 DeLTa-seq library structure

3.5 Targeted RNA-Seq Library Preparation

1. Prepare the RNA removal reaction mix as follows:

cDNA	20 μL
RNase T1	1 μL
RNase A	1 μL
Total	22 μL

2. Run the RNA removal reaction as follows:

   5 min at 37 °C

   hold at 4 °C

3. Purify RNA-depleted cDNA with 17.6 μL (0.8× volume) of AMPure XP beads according to the manufacturer's instructions. Elute with 10 μL nuclease-free water.
4. Prepare the gene-specific indexing mixture (*see* **Note 3**) as follows:

RNA-depleted cDNA	4 μL
KAPA HiFi HotStart ReadyMix	5 μL
10 μM gene-specific primer mix	1 μL
Total	10 μL

5. Run the gene-specific indexing reaction as follows:

   5 min at 95 °C

   20 s at 98 °C

   15 s at 60 °C

   40 s at 72 °C

   hold at 4 °C

6. Purify the adapter-added DNA with 8 μL (0.8× volume) of AMPure XP beads according to the manufacturer's instructions. Elute with 16 μL nuclease-free water.
7. Prepare the qPCR reaction mix as follows:

Adapter-added DNA	3.5 μL
KAPA HiFi HotStart ReadyMix	5 μL
EvaGreen dye (20× in water)	0.5 μL
10 μM second index primer	0.5 μL
10 μM PCR_P7 primer	0.5 μL
Total	10 μL

8. Run qPCR reaction as follows:

   5 min at 95 °C

   30 cycles of:

   20 s at 98 °C

   15 s at 60 °C

   40 s at 72 °C

   3 min at 72 °C

   hold at 4 °C

9. Determine the cycles at the middle of the exponential phase. This number is the optimal number of PCR cycles for library amplification.

10. Prepare the amplification reaction as follows:

Adapter-added DNA	7 μL
KAPA HiFi HotStart ReadyMix	10 μL
10 μM second index primer	1 μL
10 μM PCR_P7 primer	1 μL
Nuclease-free water	1 μL
Total	10 μL

11. Run library amplification reaction as follows:

    5 min at 95 °C

    The optimal number of cycles determined at **step 9**:

    20 s at 98 °C

    15 s at 60 °C

    40 s at 72 °C

    3 min at 72 °C

    hold at 4 °C

12. Purify the amplified library with 10 μL (1× volume) of AMPure XP beads according to the manufacturer's instructions. Elute with 15 μL of 10 mM Tris–HCl (pH 7.6).

13. Quantify the concentration of the library using the Quantus Fluorometer and the QuantiFluor ONE dsDNA System.

14. Electrophorese 1 μL of the library using a Bioanalyzer 2100 with an Agilent High Sensitivity DNA Kit.

3.6 Sequencing

Perform 150 bp paired-end sequencing of the DeLTa-seq library using an Illumina sequencing platform.

3.7 Bioinformatic Analysis of DeLTa-Seq

1. Trim low-quality bases from fastq files and append UMI information in Read2 to the read IDs of Read1 using fastp. The example command is as follows:

```
$ fastp --trim_poly_x --thread 16 --in1 <path to Read1.fq.gz>
--max_len2 1 --in2 <path to Read2.fq.gz> --out1 <file name of
output fq.gz for Read1> --out2 <file name of output fq.gz for
Read2> -l 0 -h <file name of output.html> --umi --umi_loc read2
--umi_len 10 --adapter_sequence=AGATCGGAAGAGCACACGTCTGAACTC-
CAGTCA --adapter_sequence_r2=AGATCGGAAGAGCGTCGTGTAGGGAAA-
GAGTGT.
```

2. Index a reference genome using bowtie. The example command is as follows:

```
$ bowtie-build -f <reference fasta file> <bowtie-build output
prefix name>
```

3. Map reads and sort by positions using bowtie and samtools. The example command is as follows:

```
$ bowtie --sam <bowtie-build output prefix name> <file name of
output fq.gz for Read1> | samtools view -b | samtools sort >
<file name of output bam>
```

4. Remove duplicated reads using umi_tools. The example command is as follows:

```
$ umi_tools dedup -I <file name of output bam> -S <file name of
umi-tools output bam> --per-contig -spiepr A3B2 h 0pt-umi-
separator=":".
```

5. Quantify the gene-wise and isoform-wise expression levels using RSEM. The example command is as follows:

```
$ rsem-calculate-expression –alignments <file name of
umi-tools output bam> <bowtie-build output prefix name> <RSEM
output prefix name>
```

4 Notes

1. In the first experiment, it is recommended to check the quality of the RNA in the lysate. If RNA degradation is observed, it is necessary to optimize the homogenization step.

2. Direct-lysate reverse transcription (RT) buffer successfully protects RNA from degradation in rice root lysates [14]. Thus, DeLTa-seq can be applied to rice roots.

3. Originally, we used 100 μM of gene-specific primer mix. However, we found that a lower concentration can ensure a more robust quantification of target genes. Thus, we recommend 10 μM gene-specific primer mix.

Acknowledgments

This study was supported by the Japan Science and Technology Agency (grant numbers JPMJCR15O2 and JPMJFR210B to A.J. N.). Japan Society for the Promotion of Science (JP20H00423 and JP23H00386 to A.J.N.), and Japanese Ministry of Education, Culture, Sports, Science and Technology (JP23H04967 to A.J.N.).

References

1. Hashida Y, Tezuka A, Nomura Y et al (2022) Fillable and unfillable gaps in plant transcriptome under field and controlled environments. Plant Cell Environ 45:2410–2427
2. Hirayama T, Saisho D, Matsuura T et al (2020) Life-course monitoring of endogenous phytohormone levels under field conditions reveals diversity of physiological states among barley accessions. Plant Cell Physiol 61:1438–1448
3. Satake A, Kawatsu K, Teshima K et al (2019) Field transcriptome revealed a novel relationship between nitrate transport and flowering in Japanese beech. Sci Rep 9:1–12
4. Muir P, Li S, Lou S et al (2016) The real cost of sequencing: scaling computation to keep pace with data generation. Genome Biol 17:53
5. Alpern D, Gardeux V, Russeil J et al (2019) BRB-seq: ultra-affordable high-throughput transcriptomics enabled by bulk RNA barcoding and sequencing. Genome Biol 20:71
6. Kamitani M, Kashima M, Tezuka A et al (2019) Lasy-Seq: a high-throughput library preparation method for RNA-Seq and its application in the analysis of plant responses to fluctuating temperatures. Sci Rep 9:7091
7. Li Y, Yang H, Zhang H et al (2020) Decode-seq: a practical approach to improve differential gene expression analysis. Genome Biol 21:66
8. Clark MB, Amaral PP, Schlesinger FJ et al (2011) The reality of pervasive transcription. PLoS Biol 9:e1000625
9. Chen S, Zhou Y, Chen Y et al (2018) Fastp: an ultra-fast all-in-one FASTQ preprocessor. Bioinformatics 34:i884–i890
10. Langmead B, Trapnell C, Pop M et al (2009) Ultrafast and memory-efficient alignment of short DNA sequences to the human genome. Genome Biol 10:R25
11. The SAM Format Specification (v1.4-r985) (2011)
12. Smith T, Heger A, Sudbery I (2017) UMI-tools: modelling sequencing errors in unique molecular identifiers to improve quantification accuracy. Genome Res 27: gr.209601.116
13. Zhang C, Zhang B, Lin LL et al (2013) Evaluation and comparison of computational tools for RNA-seq isoform quantification. BMC Genomics 18:1–11
14. Kashima M, Kamitani M, Nomura Y, Mori-Moriyama N, Betsuyaku S, Hirata H, Nagano AJ (2022) DeLTa-Seq: direct-lysate targeted RNA-Seq from crude tissue lysate. Plant Methods 18:99

Chapter 13

Lasy-Seq: A High-Throughput 3′ RNA-Seq Method for Large-Scale Transcriptome Profiling in Rice

Natsumi Mori-Moriyama and Atsushi J. Nagano

Abstract

Lasy-Seq (low-cost and easy RNA-seq) is a high-throughput library preparation method for 3′ RNA-seq. Usage of oligo(dT) primers with unique index sequences allows simultaneous reverse transcription (RT) reaction and sample indexing. Therefore, each post-RT step, which was performed on a sample-by-sample basis in a conventional method, can be performed on a large number of samples together in a single tube. With the Lasy-Seq, the sequencing-ready library pool can be prepared from a few to several hundred samples in a few days.

Key words Transcriptome, 3′ RNA-seq, Scalable RNA-seq, Large-scale RNA-seq, High-throughput RNA-seq

1 Introduction

The transcriptome is a global profile of total mRNA extracted from cells or tissues. In rice research, as in other species, there is a growing need for large-scale transcriptome analysis [1, 2]. Improvements in next-generation sequencing (NGS) technology have made it possible to obtain a large amount of sequence information in a single operation. Although the sequencing capacity is sufficient to perform transcriptome analysis of many samples, the throughput of library preparation for RNA-seq is limited and is considered one of the bottlenecks in large-scale transcriptome analysis.

During library preparation, mRNA is converted into cDNA fragments having sequencing adapters and sample index sequences in a multistep reaction. The conventional RNA-seq library preparation involves mRNA enrichment, reverse transcription (RT), second-strand synthesis, end-repair/dA addition, adapter ligation, size selection, and PCR amplification (*see* Fig. 1) [3]. The RNA-seq libraries are barcoded with unique index sequences to identify samples during PCR amplification. This allows the pooling of

Taiji Kawakatsu (ed.), *Rice: Methods and Protocols*, Methods in Molecular Biology, vol. 2869,
https://doi.org/10.1007/978-1-0716-4204-7_13,

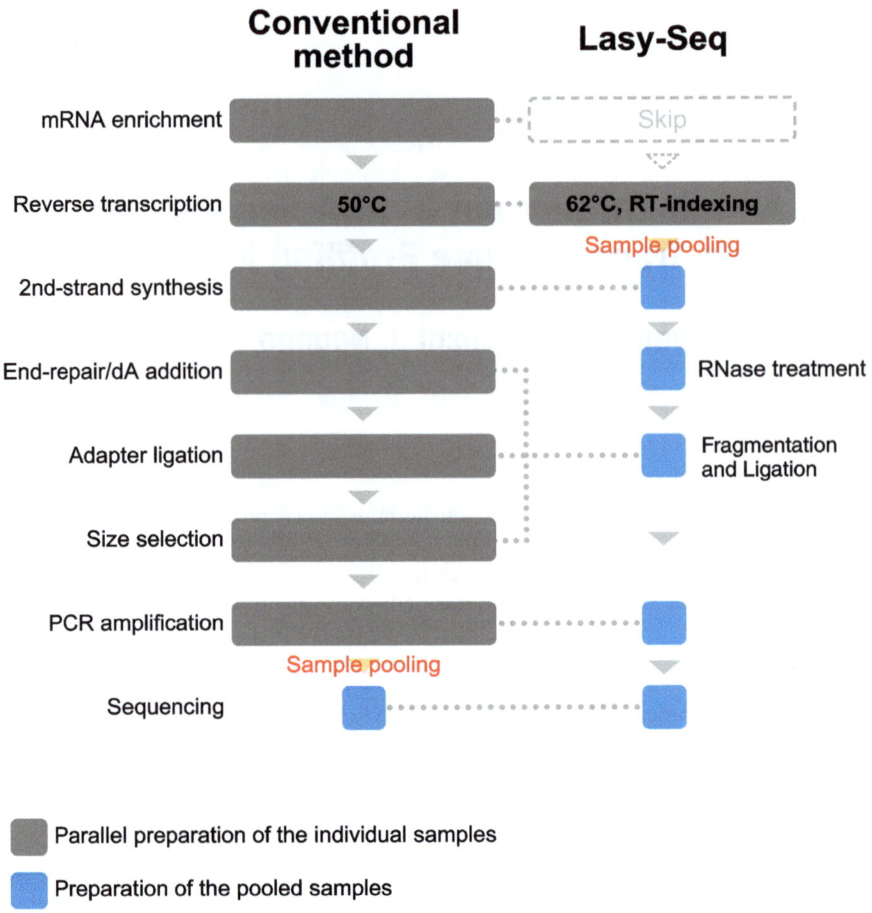

Fig. 1 Comparison of conventional method and Lasy-Seq

libraries, which are sequenced using an NGS platform. However, in this method, each step of the experimental procedure is performed for individual samples, which is costly and time-consuming, especially in large-scale projects. mRNA enrichment is closely related to RNA degradation and requires careful experimentation.

In this chapter, we describe the simplified Lasy-Seq (*l*ow-cost and *easy* RNA-*seq*), in which RT reactions and sample indexing using oligo(dT) primers with unique index sequences are performed simultaneously (*see* Fig. 1) [4]. This strategy allows for sample pooling after the RT step and post-RT step can be performed in one tube. Furthermore, high temperature during RT reaction suppresses RT from A-rich regions within rRNA so that the laborious mRNA enrichment step can be omitted. The simplified Lasy-Seq is a scalable 3′ RNA-seq library preparation from a few samples to several hundreds of samples.

2 Materials

2.1 Reverse Transcription

1. Total RNA (500 ng per sample).
2. SuperScript IV Reverse Transcriptase (Thermo Fisher Scientific) (*see* **Note 1**): This kit includes SuperScript IV Reverse Transcriptase, 5× SuperScript IV buffer, and 100 mM dithiothreitol (DTT).
3. 25 mM dNTP mix.
4. 2 μM RT primer: 5′- AGAAGACGGCATACGA GATXXXXXXXXXXGTGACTGGAGTTCA GACGTGTGCTCTTCC GATCTNNNNNNNNTTTTTTTTTTTTTTTTTTTTV-3′. “X” and “N” in the sequence indicate the 10-mer index sequences and the unique molecular identifier (UMI), respectively.
5. External RNA Controls Consortium (ERCC) RNA Spike-In control mixes.
6. Nuclease-free water.

2.2 Pooling and Purification

1. 2× Membrane Binding Solution: 4.5 M guanidine isothiocyanate, 0.5 M potassium acetate, pH 5.0.
2. Zymo-Spin I columns or equivalent silica-based spin column for DNA/RNA purification.
3. Laboratory Vacuum Manifold for silica-based spin columns.
4. Wash buffer: 100 mM Tris–HCl, pH 7.6, 80% ethanol.
5. Nuclease-free water

2.3 Second-Strand Synthesis and RNase Treatment

1. 10× Blue buffer (Enzymatics, B0110): 500 mM NaCl, 100 mM Tris–HCl, 100 mM $MgCl_2$, and 10 mM DTT, pH 7.9.
2. 2.5 mM dNTP mix.
3. 100 mM DTT.
4. RNase H (5000 U/mL).
5. DNA polymerase I (5000 U/mL).
6. Nuclease-free water.
7. RNase solution: RNase T1 (1 U/μL) and RNase A (10 mg/mL) in 50% glycerol.
8. AMPure XP (Beckman Coulter) (*see* **Note 2**).

2.4 cDNA Library Preparation and Library Amplification

1. QuantiFluor dsDNA System (Promega).
2. Quantus Fluorometer (Promega).

3. WGS Fragmentation Mix (Qiagen): This kit includes 5× WGS Fragmentation Mix and 10× Fragmentation buffer (*see* **Note 3**).
4. WGS Ligase (Qiagen): This kit includes 5× ligation buffer and WGS Ligase (*see* **Note 3**).
5. Nuclease-free water.
6. 4 mM Y-shaped adapter: 5′-A*C*C* GAGATCTACA CACTCTTTCCCTACACGACGCTCTTCCGA *T*C*T-3′ (forward adapter oligo), 5′-/5Phos/G*A*T* CGGAA GAGCGTCGTGTTAAATGTA *T*A*T-3′ (reverse adapter oligo). Asterisks (*) indicate phosphorothioate bonds. / 5Phos/indicates phosphorylation. Mix 50 μL of each 100 μM adapter oligo. Incubate at 95 °C for 2 min, then slowly cool down to 25 °C at 0.1 °C/s, and then incubate at 25 °C for 30 min. Dilute the annealed adapter (50 μM) to 4 mM with nuclease-free water and store at −20 °C. The adapter is gently thawed on ice before use (*see* **Note 4**).
7. KAPA HiFi HotStart ReadyMix (Roche; *see* **Note 5**).
8. 20× EvaGreen dye (Biotium; *see* **Note 5**).
9. Forward PCR primer: 5′-CAAGCAGAAGACGGCATACGA GAT-3′
10. Reverse PCR primer: 5′- AATGATACGGCGACCACCGA GATCTACACACTCTTTCCCT-3′ (*see* **Note 6**).
11. Real-time PCR thermal cycler.
12. AMPure XP (*see* **Note 2**).
13. Nuclease-free water.

2.5 Quality check of the Amplified Library

1. Agilent 2100 Bioanalyzer (*see* **Note 7**).
2. High-Sensitivity DNA Kit (*see* **Note 7**).
3. QuantiFluor dsDNA System (Promega).
4. Quantus Fluorometer (Promega).

3 Methods

3.1 Reverse Transcription

1. Prepare total RNA samples (50 ng/μL) in nuclease-free water.
2. Prepare the reverse transcription (RT) reaction master mix on ice as follows:

5× Superscript IV buffer	4.0 μL
100 mM DTT	2.0 μL
25 mM dNTP mix	0.4 μL

(continued)

Superscript IV reverse transcriptase	0.1 μL
Nuclease-free water	2.5 μL
Total	9 μL

If you want to assess occurrences of index hopping (*see* **Note 8**), in some samples, the master mix is reduced to 1 μL of nuclease-free water and replaced with 1 μL of ERCC RNA Spike-In control mixes diluted 1/100.

3. Dispense 9 μL of RT reaction master mix into PCR tubes or 96-well plates on ice.
4. Add 10 μL of 50 ng/μL RNA.
5. Add 1 μL of 2 μM RT primer with index sequences to each sample (*see* **Note 9**).
6. Run the RT reaction as follows:

   10 min at 65 °C

   15 min at 80 °C

   hold at 4 °C

3.2 Pooling and Purification

1. Pool RT products from **step 6** in Subheading 3.1 (*see* **Note 10**).
2. Add an equal volume of 2× Membrane binding solution and mix gently.
3. Set a Zymo-Spin I column on the Laboratory Vacuum Manifold.
4. Add 700 μL of the pooled RT product from **step 2** to a Zymo-Spin I column.
5. Incubate at room temperature for 1 min.
6. Apply vacuum and bind the pooled RT product to silica membrane of the Zymo-Spin I column.
7. Add the remaining pooled RT product before the column becomes empty.
8. Apply 700 μL of wash buffer.
9. Apply vacuum and wash the column.
10. Repeat **steps 8** and **9**.
11. Place the column on a new 2.0 mL tube and centrifuge at 16,000× *g* for 2 min to remove residual ethanol.
12. Transfer the column on a new tube.
13. Add 100 μL of nuclease-free water.
14. Incubate for 1 min.
15. Centrifuge at 16,000× *g* for 2 min and elute the pooled RT product.

3.3 Second-Strand Synthesis and RNase Treatment

1. Prepare the second-strand synthesis reaction mix on ice as follows:

10× blue buffer	4 μL
2.5 mM dNTP mix	2 μL
100 mM DTT	1 μL
Pooled RT products from **step 15** in Subheading 3.2	30 μL
RNase H (5000 U/mL)	1 μL
DNA polymerase I (5000 U/mL)	2 μL
Total	40 μL

Store the remaining 70 μL of pooled RT products at −20 °C.

2. Run second-strand synthesis reaction as follows:
   2 h at 16 °C
   hold at 4 °C
3. Add 2 μL of RNase solution.
4. Incubate at 37 °C for 5 min.
5. Add 33.6 μL of AMPure XP (0.8× volume) and mix well by vortexing or pipetting.
6. Incubate for 6 min at room temperature.
7. Place the tube on a magnetic stand for 3 min.
8. Discard the supernatant.
9. Add 500 μL of 75% ethanol.
10. Incubate for 30 s.
11. Discard the supernatant.
12. Repeat **steps 9–11**.
13. Briefly spin down and place the tube on a magnetic stand.
14. Remove as much ethanol as possible.
15. Keep the lid open and air-dry at room temperature for 2 min on a magnetic stand.
16. Remove the tube from a magnet stand.
17. Add 10 μL of nuclease-free water and mix well.
18. Incubate at room temperature for 5 min.
19. Place the tube on a magnetic stand for 2 min.
20. Transfer the supernatant to a new tube.

3.4 cDNA Library Preparation

1. Quantify the accurate amount of double-stranded cDNA using QuantiFluor dsDNA System and Quantus Fluorometer, following the manufacturer's protocol.

2. Prepare the fragmentation reaction mix on ice as follows:

10× fragmentation buffer	2.5 μL
5× WGS fragmentation mix	5.0 μL
Double-stranded cDNA	50 ng (*see* **Note 11**)
Nuclease-free water	Up to 25.0 μL
Total	25.0 μL

This step performs fragmentation, end-repair, and dA-tailing in a single tube. Store the remaining double-stranded cDNA sample at −20 °C.

3. Run the fragmentation reaction as follows:

   1 min at 4 °C

   7 min at 32 °C

   30 min at 65 °C

   hold at 4 °C

4. Prepare the adapter ligation reaction on ice as follows:

Fragmented and end-prepped cDNA from **step 3**	25 μL
4 mM Y-shaped adapter	2 μL
5× ligation buffer	10 μL
WGS ligase	5 μL
Nuclease-free water	8 μL
Total	50 μL

5. Run the adapter ligation reaction with heated lid set at 65 °C as follows:

   1 min at 4 °C

   15 min at 20 °C

   hold at 4 °C

6. Repeat **steps 5–20** in Subheading 3.3 with 40 μL of AMPure XP and elute in 50 μL of nuclease-free water.
7. Repeat **steps 5–20** in Subheading 3.3 with 40 μL of AMPure XP to completely remove the Y-shaped adapter and elute in 17 μL of nuclease-free water.

3.5 Library Amplification

First, determine the optimal number of PCR cycles using real-time PCR to prevent excessive duplication of library sequences. Second, amplify the adapter-ligated cDNA with the determined optimal number of PCR cycles to complete the library.

1. Prepare the qPCR reaction mix as follows:

Adapter-ligated DNA from **step 7** in Subheading 3.4	3.5 μL (approximately 10 ng)
KAPA HiFi HotStart ReadyMix	5 μL
20× EvaGreen dye	0.5 μL
Forward PCR primer	0.5 μL
Reverse PCR primer	0.5 μL
Total	10 μL

2. Run the qPCR reaction as follows:

   5 min at 95 °C

   30 cycles of:

   20 s at 98 °C

   15 s at 60 °C

   40 s at 72 °C

   3 min at 72 °C

   hold at 4 °C

3. Determine the cycles at the middle of the exponential phase (*see* Fig. 2). This number is the optimal number of PCR cycles for library amplification.
4. Prepare the library amplification reaction as follows:

Adapter-ligated DNA from **step 7** in Subheading 3.4	12 μL (approximately 35 ng)
KAPA HiFi HotStart ReadyMix	15 μL
Forward PCR primer	1.5 μL
Reverse PCR primer	1.5 μL
Total	30 μL

5. Run the library amplification reaction as follows:

   5 min at 95 °C

   The optimal number of cycles determined at **step 3**:

   20 s at 98 °C

   15 s at 60 °C

   40 s at 72 °C

   3 min at 72 °C

   hold at 4 °C

6. Repeat **steps 5–20** in Subheading 3.3 with 30 μL of AMPure XP and elute in 15 μL of nuclease-free water.

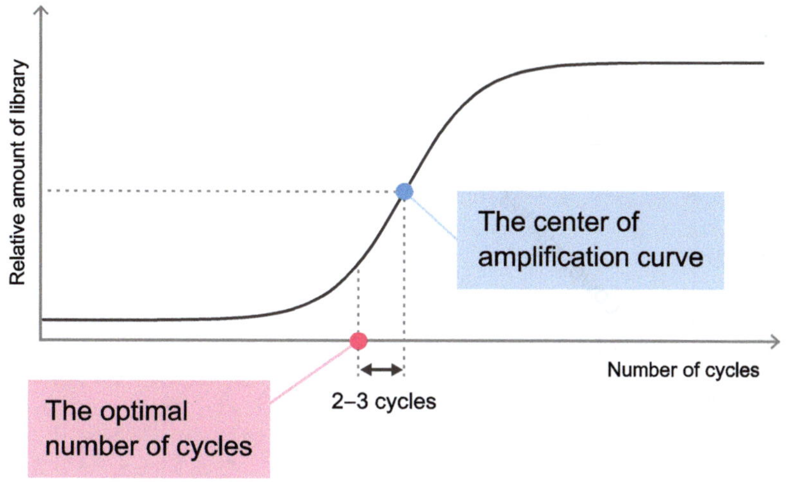

Fig. 2 Optimization of the number of PCR cycles using real-time PCR

3.6 Quality Check of the Amplified Library

1. Quantify the amount of amplified library using QuantiFluor dsDNA System and Quantus Fluorometer, following the manufacturer's protocol.
2. Assess size distribution of amplified library using a DNA High Sensitivity Kit and an Agilent 2100 Bioanalyzer, following the manufacturer's protocol. The expected size distribution ranges from 200 bp to 1500 bp with an average size of 450–500 bp and no shape peaks of a specific length (*see* Fig. 3a). Perform smear assay to quantify the library concentration. The library can be sequenced on the Illumina platform (*see* Fig. 3b and **Note 12**).

4 Notes

1. In Lasy-Seq, the RT reaction temperature is 65 °C for mRNA-specific RT. Therefore, we recommend SuperScript IV Reverse Transcriptase that has high thermostability and inhibitor tolerance.
2. We recommend AMPure XP as a magnetic bead for nucleic acid purification due to its high yield. Other similar products may be used, but input amounts and purification procedures should be adjusted according to the product manual.
3. The latest Lasy-Seq protocol is optimized to use WGS Fragmentation Mix and WGS Ligase for the fragmentation, end-repair, dA addition, and ligation steps. If other reagents are used, reaction conditions must be considered. Please see [4, 5] for the old protocol using tagmentation by Nextera transposase.

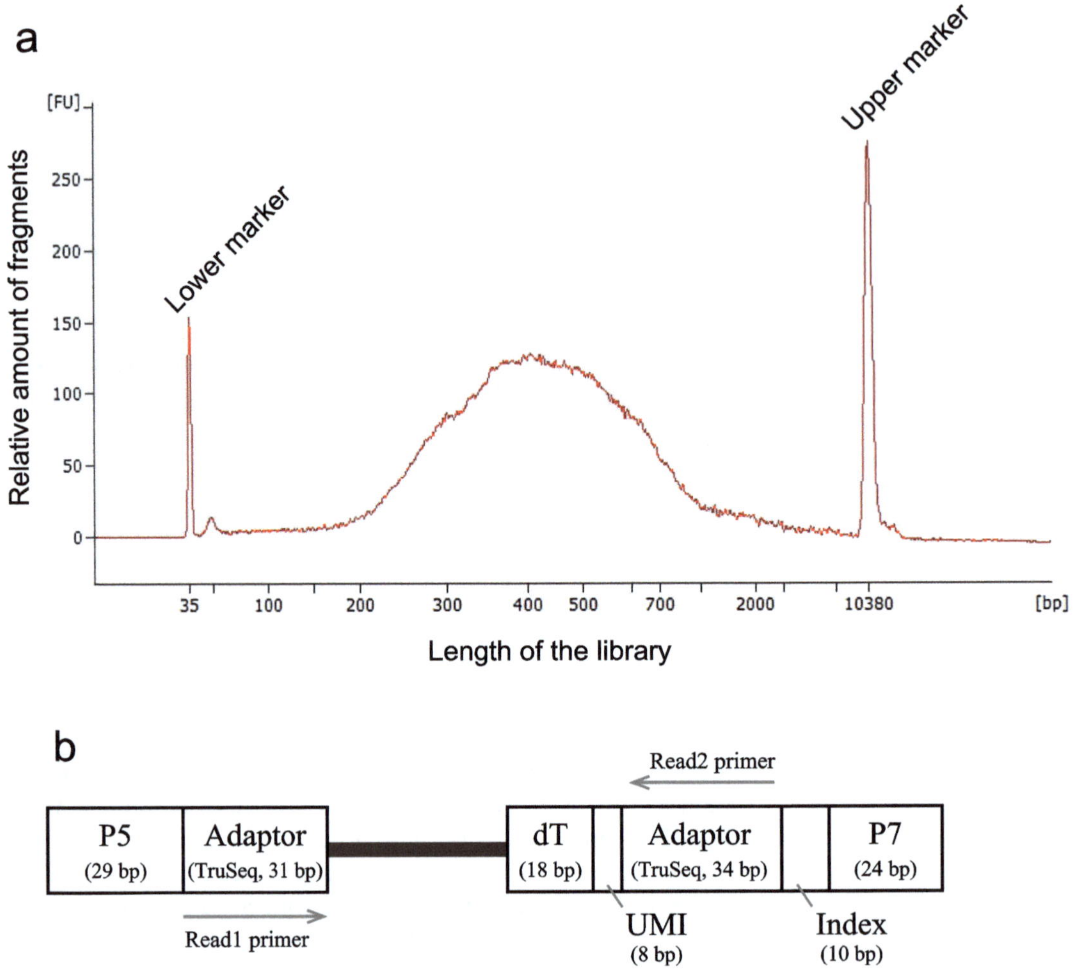

Fig. 3 Quality check of library. (**a**) Representative of recommended library distribution. (**b**) The structure of the Lasy-Seq library

4. The Y-shaped adapter must be pre-annealed with partially complementary oligo DNAs. A mixture of 100 mM forward and reverse adapter oligos is annealed using a thermal cycler at 95 °C for 2 min, which is then slowed down to 25 °C at 0.1 °C/s and then 25 °C for 30 min. The annealed adapter (50 μM) is diluted to 4 mM in nuclease-free water and stored at −20 °C. The adapter is gently thawed on ice before use.
5. We recommend that the reaction conditions for real-time PCR and subsequent PCR be as similar as possible to determine the exact optimal number of cycles. We use the same high-fidelity PCR enzymes and a reaction mixture supplemented with DNA-specific dyes for real-time PCR.
6. Using the reverse PCR primer with the second index sequence, more samples can be pooled and sequenced, for example, 5′-

AATGATACGGCGACCACCGAGATCTACACXXXXXXXXXXACACTCTTTCCCTACACGA-3′. "X" in the sequence indicates the 10-mer index sequences.

7. It is recommended that a completed library be quality-checked for size distribution and concentration using the Agilent Bioanalyzer or similar instruments before sequencing.
8. Index hopping, the exchange of index sequences between samples, occasionally occurs during library preparation and/or sequence reactions on the Illumina platform [6].
9. Lasy-Seq performs sample indexing using RT primers with a 10-mer index sequence in the RT step. To identify individual samples, RT primers with one index sequence per RNA sample were used.
10. The pooling method was used according to [7] and discussed in detail. Set a metal frame (special order product, stainless steel, height 42 mm, width 127 mm, depth 84.5 mm) on a one-well reservoir (Nalgene Disposable Polypropylene Robotic Reservoirs, 1200–1301, Thermo Fisher Scientific), and cover the four sides with Parafilm. Place the RT reaction plate upside down on a one-well reservoir with a metal frame. Fill up the gap between an RT reaction plate and a one-well reservoir using a paraffin film to prevent leakage of RT products. Collect RT products into a one-well reservoir via centrifugation at 1500 × *g* for 3 min.
11. We recommend an input amount of at least 50 ng of double-stranded cDNA for fragmentation. Inputting small amounts of double-stranded cDNA can result in shorter fragments.
12. The first reads are used for mapping based on the principle of Lasy-Seq. However, the second read contains UMI and polyT; therefore, the first ten bases of this read can be cut out as UMI.

Acknowledgments

This study was supported by the Japan Science and Technology Agency (grant numbers JPMJCR15O2 and JPMJFR210B to A.J. N.), Japan Society for the Promotion of Science (JP20H00423 and JP23H00386 to A.J.N.), and Japanese Ministry of Education, Culture, Sports, Science and Technology (JP23H04967 to A.J.N.).

References

1. Hashida Y, Tezuka A, Nomura Y et al (2022) Fillable and unfillable gaps in plant transcriptome under field and controlled environments. Plant Cell Environ 45:2410–2427
2. Kashima M, Sakamoto RL, Saito H et al (2021) Genomic basis of transcriptome dynamics in rice under field conditions. Plant Cell Physiol 62:1436–1445

3. Ishikawa T, Kashima M, Nagano AJ et al (2017) Unfolded protein response transducer IRE1-mediated signaling independent of XBP1 mRNA splicing is not required for growth and development of medaka fish. elife 6:1–29
4. Kamitani M, Kashima M, Tezuka A et al (2019) Lasy-Seq: a high-throughput library preparation method for RNA-Seq and its application in the analysis of plant responses to fluctuating temperatures. Sci Rep 9:1–14
5. Nagano AJ (2018) Lasy-Seq protocol. https://sites.google.com/view/lasy-seq/
6. Illumina Inc (2017) Effects of index misassignment on multiplexing and downstream analysis. https://www.illumina.com/content/dam/illumina-marketing/documents/products/whitepapers/index-hopping-white-paper-770-2017-004.pdf. Accessed 24 Jun 2024
7. Sasagawa Y, Danno H, Takada H et al (2018) Quartz-Seq2: a high-throughput single-cell RNA-sequencing method that effectively uses limited sequence reads. Genome Biol 19:1–24

Chapter 14

Profiling of 24-nt phasiRNAs in Rice Anther by Massively Parallel Sequencing and Stem-Loop Reverse Transcription PCR

Huong Ta and Ken-Ichi Nonomura

Abstract

Small RNAs are highly abundant and play important roles in plant reproduction. Profiling of small RNAs in reproductive tissues is a critical step in understanding their biology. Here, we describe a protocol for small RNA profiling in rice anthers, with a focus on an abundantly expressed but little-understood reproductive small RNA class named 24-nucleotide phased secondary small interfering RNAs (24-nt phasiRNAs). The protocol details small RNA library preparation steps for high-throughput sequencing, with a subsequent bioinformatic analysis framework. A low-throughput stem-loop reverse transcription PCR protocol for quick semi-quantification of small RNAs is also included as a complementary approach. The collective protocol, although focused on 24-nt phasiRNAs, should be amenable to other small RNA classes in various plant reproductive tissues.

Key words 24-nt phasiRNAs, Reproduction, Small RNAs, Stem-loop RT-PCR

1 Introduction

Plant reproduction is critical both in nature and agriculture. Successful plant reproduction requires concerted contributions of multiple regulatory factors including various small RNA species. Two largest and most well-studied classes of small RNAs are microRNAs and siRNAs, which are mostly 21–22 nt and 24 nt in length, respectively. These small RNAs associate with different members of Argonaute (AGO) protein family to exert regulatory roles through transcriptional, post-transcriptional, and/or translational control of gene expression.

Twenty-four-nucleotide phasiRNAs represent a novel and less-known class of small RNAs, which have been discovered in the reproductive stage of both monocotyledonous and dicotyledonous plants [1, 2]. Rice meiotic anthers, starting from early meiosis throughout to tetrad stages, are enriched with 24-nt phasiRNAs.

Taiji Kawakatsu (ed.), *Rice: Methods and Protocols*, Methods in Molecular Biology, vol. 2869,
https://doi.org/10.1007/978-1-0716-4204-7_14,

These phasiRNAs are produced from *24-PHAS* loci, which are several hundred nucleotides in length and are transcribed under the regulation of transcription factors such as ETERNAL TAPETUM1 [3]. The transcripts are then double-stranded thanks to a trigger by microRNA 2275 and the activity of RNA-DEPENDENT RNA POLYMERASE 6 protein, before being diced into precise head-to-tail 24-nt fragments by DICER-LIKE 5 [4, 5]. The interacting AGO protein(s), targets, functions, and mechanism of action of 24-nt phasiRNAs remain elusive.

Profiling of 24-nt phasiRNAs or any other small RNAs in reproductive tissues is the first important step in achieving understanding of their biology. Comprehensive transcriptome-wide profiling of small RNAs can be obtained using massively parallel sequencing, also known as next-generation sequencing (NGS) technologies. Library preparation for NGS involves adaptor ligation, synthesis of the complementary DNA, and PCR enrichment with appropriate barcoding that allows parallel processing of multiple samples. Commercial kits have also been made available to streamline the procedures of small RNA profiling using various sequencing platforms such as Illumina Miseq and NovaSeq. Depending on the specific research interest, different small RNA classes can be selected, sequenced, and annotated from a single NGS run, using experimental and/or computational tools. For example, 24-nt phasiRNAs can be enriched by size selection on a PAGE gel before sequencing and annotated using customized pipelines or recent software specifically designed for phased siRNAs such as *PHASIS* [6].

On the other hand, NGS can be costly and time-consuming, whereas transcriptome-wide profiling of small RNAs is not always necessary. In addition, an orthogonal method to validate NGS data is required in some cases. Stem-loop reverse transcription (RT)-PCR offers a simple and low-cost alternative to quickly profile, with lower throughput, small RNAs of interest. We present in this chapter detailed steps to profile 24-nt phasiRNAs from rice anther samples using both NGS and stem-loop RT-PCR.

2 Materials

2.1 Small RNA Library Preparation for Illumina Sequencing Platforms

1. Anthers from rice plant: Sample rice anthers according to length as listed in Table 1. Snap-freeze samples using liquid nitrogen, and store at −80 °C. We use a minimum of 90 Stage 1 (ST1) or 60 Stage 4 (ST4) anthers (15 and 10 florets, respectively).
2. Tissue homogenization equipment such as mortar and pestle or bead mill homogenizer.
3. TRIzol reagent.

Table 1
Rice (*Oryza sativa* ssp. *japonica* cv. Nipponbare) anther length and developmental stages for sampling

Stage	Anther size (mm)	Anther stage
ST1	0.3–0.4	Pre-meiosis
ST2	0.4–0.5	Early meiosis
ST4	0.8–0.9	Tetrad

4. Chloroform.
5. Isopropyl alcohol.
6. 75% ethanol.
7. Nucleospin RNA XS kit (Macherey-Nagel) or alternative reagents for DNase I treatment and recovery of RNAs.
8. NEBNext Multiplex Small RNA Library Prep Set for Illumina Set 1 (NEB) or other small RNA library prep kits (*see* **Note 1**).
9. 6% PAGE gel.
10. SYBR Gold nucleic acid gel stain.
11. RNase-free water.

2.2 Stem-Loop RT-PCR

1. Stem-loop RT primer: 5′-GTCGTATCCAGTGCAGGGT CCGAGGTATTCGCACTGGATACGACNNNNNN-3′, with NNNNNN being the reverse complement of the target RNA's last six nucleotides.
2. Forward PCR primer: 5′-GCGGCGGNNNNNNNNNNN NNNNNN-3′, with NNNNNNNNNNNNNNNNNN being the target RNA sequence excluding the last six nucleotides matching stem-loop RT primer.
3. Universal reverse PCR primer: 5′-GTGCAGGGTCCGAGGT-3′.
4. Superscript III Reverse Transcriptase kit (Invitrogen).
5. Low molecular weight ladder.
6. 0.5× TAE: 20 mM Tris–HCl, 10 mM acetic acid, 0.5 mM EDTA.
7. Agarose gel: 4% agarose in 0.5× TAE.
8. 2× GoTaq Green Master Mix (Promega).

2.3 Bioinformatic Analysis

1. Trim Galore (https://github.com/FelixKrueger/TrimGalore) or Trimmomatic (https://github.com/usadellab/Trimmomatic) [7].
2. FastQC (https://www.bioinformatics.babraham.ac.uk/projects/fastqc/).

3. *PHASIS* (https://github.com/atulkakrana/PHASIS) [6].
4. ShortStack (https://github.com/MikeAxtell/ShortStack) [8] or BWA (https://github.com/lh3/bwa) [9].
5. SAMtools (https://www.htslib.org/) [10].
6. BEDtools (https://bedtools.readthedocs.io/) [11].

3 Methods

An overview of our small RNA profiling strategy is presented in Fig. 1. We detail in the following subheadings the steps for total RNA preparation, small RNA library construction to run on Illumina short-read sequencing platforms, small RNA-seq analysis, and small RNA stem-loop RT-PCR.

3.1 Isolation of Total RNAs from Rice Anther

To obtain high and consistent quality of anther samples from different reproductive stages, we routinely isolate total RNAs in two steps: (i) isolation with TRIzol reagent followed by (ii) on-column DNase I treatment and clean-up. The resultant total RNA can be subjected directly to small RNA library construction for NGS and stem-loop RT-PCR.

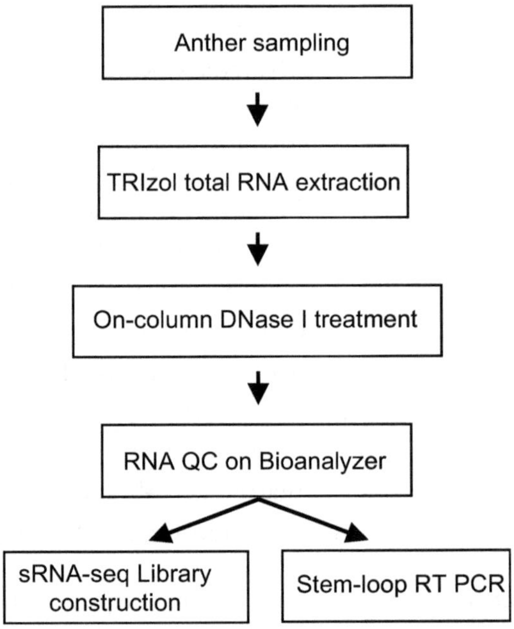

Fig. 1 Overview of our small RNA profiling strategy. Total RNAs are extracted from rice anther samples using TRIzol, purified, and QC-ed. The purified total RNAs are subject to small RNA-seq and stem-loop RT-PCR

3.1.1 Total RNA Isolation

1. Grind snap-frozen anthers using homogenization equipment.
2. Add 1 mL of TRIzol reagent to homogenized tissues, mix thoroughly, and incubate at room temperature for 5 min.
3. Add 0.2 mL of chloroform, vortex for 10 s, and incubate for 3 min.
4. Centrifuge at 12,000× *g* for 15 min at 4 °C.
5. Transfer the aqueous phase to a fresh 1.5 mL tube.
6. Add 0.5 mL of isopropyl alcohol, vortex briefly, and incubate for 10 min at room temperature.
7. Centrifuge at 12,000× *g* for at least 15 min at 4 °C (*see* **Note 2**).
8. Remove supernatant and wash the pellet with 1 mL of 75% ethanol.
9. Air-dry and resuspend the RNA pellet in 100 μL of RNase-free water.

3.1.2 DNase I Treatment and Clean-Up

The following steps describe DNase I treatment and clean-up using Nucleospin RNA XS kit. Alternative procedures for DNase I treatment and recovery of RNAs could be used.

1. Add 100 μL of RA1/ethanol mix (25 μL of buffer RA1 + 75 μL of 100% EtOH) to 100 μL of TRIzol-isolated total RNA. Mix and load onto NucleoSpin Column (blue ring).
2. Centrifuge at 11,000× *g* for 30 s.
3. Load 25 μL of rDNase reaction mix (1 rDNase/9 rDNase reaction buffer) onto the NucleoSpin Column. Incubate for 45 min at room temperature.
4. Add 100 μL of RA2 buffer, and incubate for 2 min at room temperature to inactivate the rDNase.
5. Add 400 μL of RA3 buffer and centrifuge at 11,000× *g* for 30 s.
6. Repeat **step 5** with 200 μL of RA3 buffer.
7. Dry column by centrifugation at 11,000× *g* for 2 min.
8. Add 16 μL RNase-free water and centrifuge at 11,000× *g* for 1 min to elute the RNA.
9. Check RNA quality using Agilent 2100 Bioanalyzer and RNA 6000 Pico Kit (*see* **Note 3**).

3.2 Massively Parallel Sequencing of Small RNA Library

3.2.1 Small RNA-Seq Library Preparation

1. Prepare small RNA sequencing library from 100–1000 ng of total RNAs using appropriate kit (*see* **Notes 1** and **4**).
2. Perform size selection and recovery of small RNA fraction of interest. For size selection of 24-nt fraction, we ran gel electrophoresis for 1 h at 120 V using 6% PAGE gels, followed by staining with SYBR Gold and excision of 146-nt bands (*see* Fig. 2a and **Note 5**).

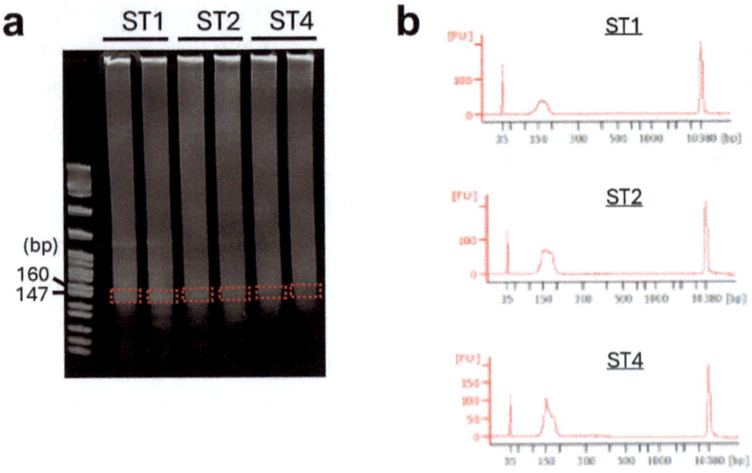

Fig. 2 Small RNA-seq library prep from rice anther total RNAs. (**a**) Amplified library resolved on 6% PAGE gel. Bands corresponding to original 24-nt RNA fraction, ~146 nt at completion of library construction including adaptors and a barcode, are excised (red dashed boxes). (**b**) Quality control of library on Agilent 2100 Bioanalyzer. In detailed report produced by Bioanalyzer, peaks of ~150 nt get clearer in ST2 and ST4 (*see* **Note 6**)

3. Check library quality using Agilent 2100 Bioanalyzer and High Sensitivity DNA Analysis kit (*see* Fig. 2b and **Note 6**).

3.2.2 Bioinformatic Analysis of 24-nt phasiRNAs

Our bioinformatic procedure for 24-nt phasiRNA analysis is summarized in Fig. 3a. We use two strategies to extract 24-nt phasiRNAs: (i) *PHASIS* is a command-line tool designed for calling of phased small interfering RNAs [7]. This tool is useful especially in identifying new *24-PHAS* loci when the coverage and read depth are good. However, depending on the nature of biological samples and sequencing libraries, 24-nt phasiRNA read depth in certain loci may be low. This may affect phasing score calculation and result in the discard of bona fide 24-nt phasiRNAs. Hence, (ii) using standard bioinformatic tools (detailed below) to identify potential 24-nt phasiRNA reads that fall into known *24-PHAS* loci [1, 3] is an essential complementary approach. This latter approach is readily amenable to the analysis of other small RNA classes.

1. Trim reads to remove adapters and low-quality reads/nucleotides using Trim Galore or Trimmomatic [7].
2. Quality check post-trimming reads using FastQC. Repeat the trimming step if necessary.
3. Perform de novo identification of 24-nt phasiRNA clusters using *PHASIS* on trimmed reads [6]. Refer to the tool's official manual for detailed operation instructions.

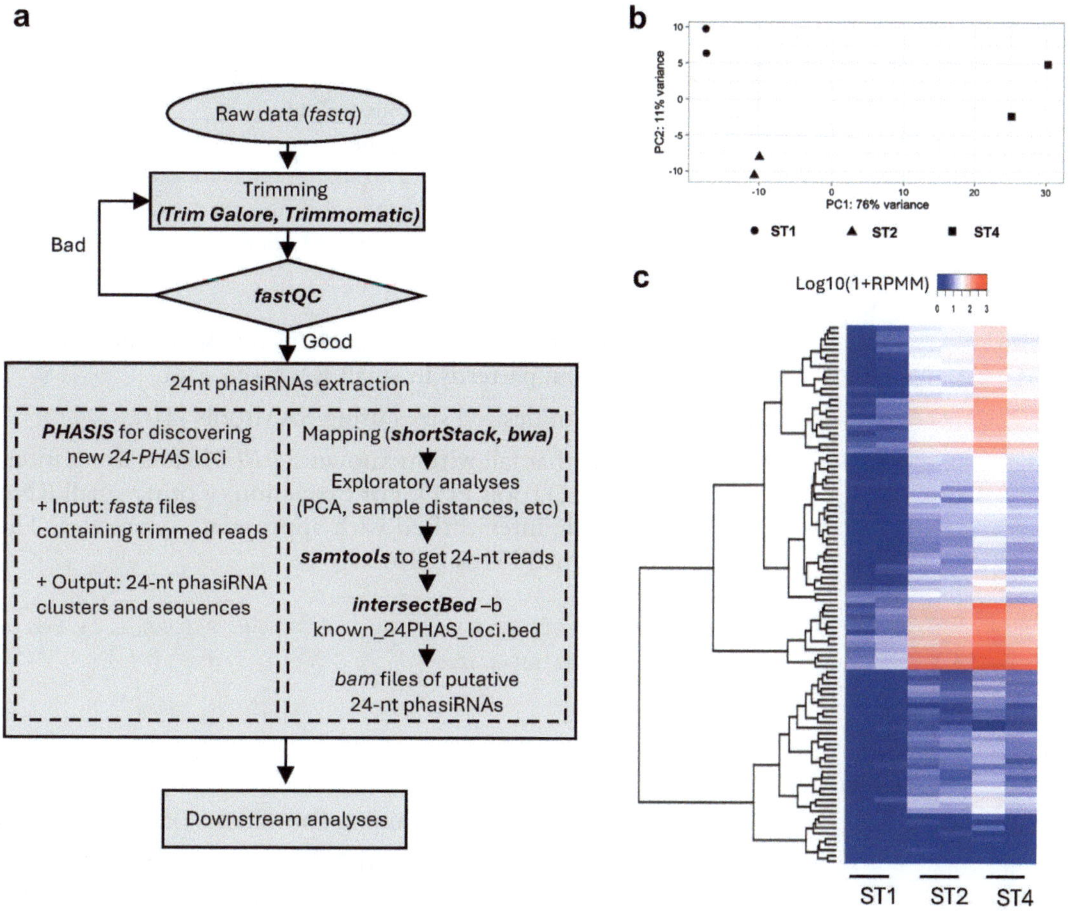

Fig. 3 Bioinformatic analysis of 24-nt phasiRNAs. (**a**) General workflow for processing small RNA-seq data and extracting 24-nt phasiRNAs. (**b**) PCA analysis of small RNAs of selected sizes in three anther stages. Plot was produced from BWA alignment files. (**c**) Clustering heatmap of 24-nt phasiRNAs from rice anther samples

4. In parallel, proceed with the following steps to extract 24-nt phasiRNAs from known loci.
5. Align trimmed reads from **step 2** to a reference genome using ShortStack [8] or BWA [9] (*see* **Note 7**). The example command using ShortStack is as follows:

```
$ ShortStack \
 --mismatches 0 \
 --outdir output_name \
 --readfile trimmed_Read.fq \
 --dicermin 19 \
 --dicermax 35 \
 --pad 100 \
 --genomefile genome.fa
```

Suggested options:

```
--mismatches: number of mismatches allowed
--outdir: directory to write output to
--readfile: file(s) to be aligned
--dicermin: Minimum size of a Dicer-processed small RNA
--dicermax: Maximum size of a Dicer-processed small RNA
--pad [distance]: merge clusters of small RNAs within the
provided distance
```

6. Perform exploratory analyses such as PCA (*see* Fig. 3b) to examine general patterns in small RNA-seq data.
7. Extract reads of desired sizes using SAMtools [10].
8. Extract reads that fall within known *24-PHAS* loci using intersectBed in BEDTools [11]. For extraction of other small RNA classes, provide intersectBed with appropriate references. The example command is as follows:

```
intersectBed -a aligned_24nt.bam -b known_24PHAS_loci.bed >
putative_24nt_phasiRNAs.bam
```

Options:

-a: bam file to extract from

-b: reference bed file(s) containing coordinates of loci of interest

9. Use the identified 24-nt phasiRNA reads in customized analyses. We exemplify here a clustering heatmap (*see* Fig. 3c).

3.3 Stem-Loop Reverse Transcription PCR

This section describes stem-loop RT-PCR as a complementary/alternative approach for NGS profiling of small RNAs. In stem-loop RT-PCR, a stem-loop RT primer is designed to complement a short (6-nt) part of the 3′ terminus of the target small RNA, without allowing binding of longer sequences, i.e., precursor transcript or genomic DNA contamination (*see* Fig. 4a). Reverse transcription is carried out to make the full complementary strand before an end-point (described in this protocol) or real-time quantitative PCR [12, 13] can be performed. Primers used in these PCRs include a forward primer containing the rest of the target small RNA sequence and a universal reverse primer, which is part of the stem-loop RT primer sequence. Multiple stem-loop RT primers can be used simultaneously, increasing the number of targets that can be analyzed from a single RT reaction.

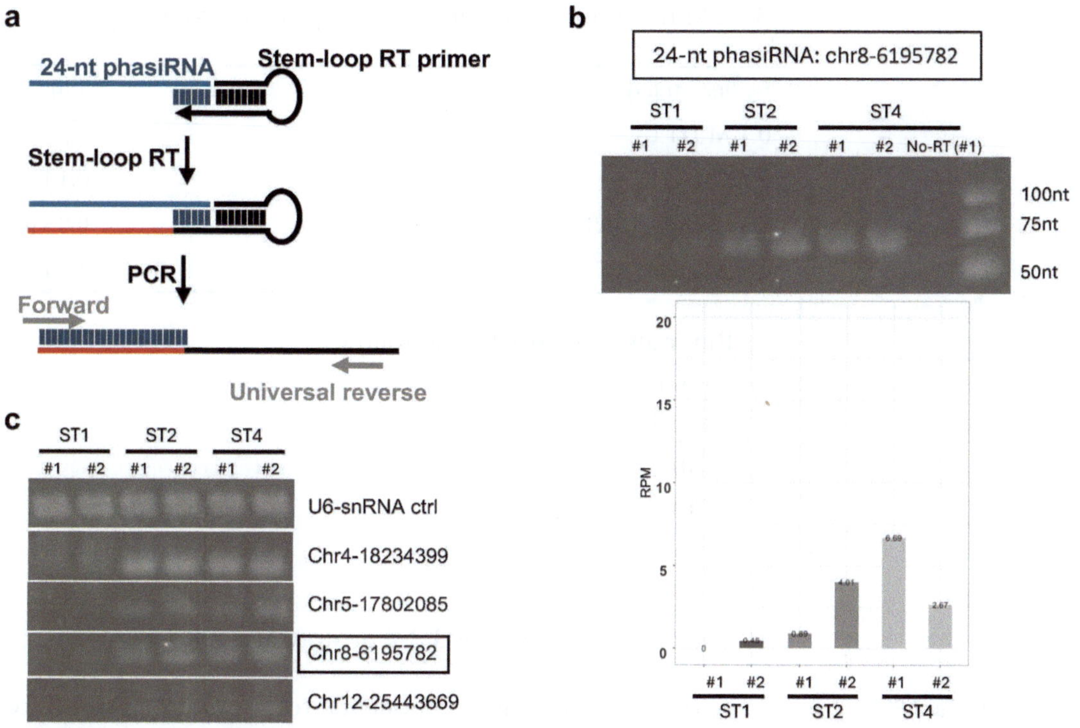

Fig. 4 24-nt phasiRNA detection and semi-quantification using stem-loop RT-PCR. (**a**) Schematic of stem-loop RT-PCR design principle. (Adapted from Varkonyi-Gasic et al. [13]). (**b**) Example of stem-loop RT end-point PCR product of a 24-nt phasiRNA on chromosome 8 resolved on 4% agarose gel (upper) versus read counts from NGS sequencing (lower) from the same total RNA sample. (**c**) Multiplexed stem-loop RT-PCR for multiple 24-nt phasiRNAs. The 24-nt phasiRNA on chromosome 8 (boxed) is also shown in (**b**)

3.3.1 Stem-Loop RT

1. Add the following components to a sterile nuclease-free PCR tube:

10 mM dNTP mix	0.5 μL
Stem-loop RT primers (10 μM)	0.1 μL each
Total RNA	100–500 ng
Nuclease-free water	Up to 13.65 μL
Total volume	13.65 μL

2. Run assembled reaction as follows:

   5 min at 65 °C

   2 min on ice

   Hold at 4 °C

3. Add the following components to product from **step 2**:

5× first-strand buffer	4 μL
0.1 M DTT	2 μL
RNaseOUT	0.1 μL
Superscript III RT	0.25 μL
Total volume	20 μL

4. Run pulsed RT reaction as follows:

   30 min at 16 °C

   60 cycles of:

   30 s at 30 °C

   30 s at 42 °C

   1 s at 50 °C

   5 min at 85 °C

5. Proceed to Subheading 3.3.2 or store at −20 °C.

3.3.2 End-Point PCR

1. Add the following components to a sterile nuclease-free PCR tube:

2× GoTaq green master mix	5 μL
Forward primer (10 μM)	0.2 μL
Universal reverse primer (10 μM)	0.2 μL
Stem-loop RT product from **step 4** in Subheading 3.3.1	1 μL
Nuclease-free water	3.6 μL
Total volume	10 μL

2. Pre-heat thermal cycler to 94 °C and run end-point PCR with the following conditions:

   2 min at 94 °C

   35 cycles of:

   15 s at 94 °C

   1 min at 60 °C

   Hold at 16 °C

3. Electrophorese PCR products on a 4% agarose gel in 0.5× TAE buffer at 100 V for 50 min.

4 Notes

1. Libraries prepared with the NEBNext Multiplex Small RNA Library Prep Set are single indexed, which is compatible with MiSeq but is not recommended to be run on the NovaSeq platform due to index hopping. If sequencing is performed using the NovaSeq platform, use small library preparation kit with unique dual index.
2. To increase the total RNA yield, we centrifuge the mixture at 12,000× *g* for 60 min as compared to 15 min in the standard protocol.
3. Generally, RNA samples with a RIN of 7 or above are considered of good quality. However, plant RNA samples can give poor RIN or no RIN on Agilent 2100 Bioanalyzer [14]. We noticed that the pre-meiosis anther samples usually contain very high amounts of shorter RNA fragments, presumably small RNA precursors, which may result in no or poor RIN value when a high amount of total RNA is loaded.
4. The NEBNext Multiplex Small RNA Library Prep Set expects 100 ng–1 μg of total RNAs. Higher range of the amounts does not appear to significantly increase the quality of the library. We used 15 cycles for final library amplification.
5. We always experienced a smearing pattern on 6% PAGE gel of libraries from rice anther samples, as opposed to distinct bands of animal examples given in the NEBNext Multiplex Small RNA Library Prep Set manual. Simply excise the region corresponding to the expected size. To recover sRNAs from the gel, we followed steps as stated in the NEBNext Multiplex Small RNA Library Prep Set standard protocol.
6. The expected peak for original 24-nt RNA fragments is approximately 146 nt including adaptors and a barcode. According to the NEBNext Multiplex Small RNA Library Prep Set manual: "The bands on the Bionalyzer electropherograms resolve in sizes ~6–8 nucleotides larger than sizes observed on PAGE gels and can shift from sample to sample due to an incorrect identification of the marker by the bioanalyzer software."
7. Rice genome and other references can be downloaded from the Rice Annotation Project Database (RAP-DB) at https://rapdb.dna.affrc.go.jp/download/irgsp1.html.

Acknowledgments

This work is supported by JSPS KAKENHI grant number 21H04729 to KIN.

References

1. Johnson C, Kasprzewska A, Tennessen K et al (2009) Clusters and superclusters of phased small RNAs in the developing inflorescence of rice. Genome Res 19:1429–1440
2. Xia R, Chen C, Pokhrel S et al (2019) 24-nt reproductive phasiRNAs are broadly present in angiosperms. Nat Commun 10:627
3. Ono S, Liu H, Tsuda K et al (2018) EAT1 transcription factor, a non-cell-autonomous regulator of pollen production, activates meiotic small RNA biogenesis in rice anther tapetum. PLoS Genet 14:e1007238
4. Song X, Li P, Zhai J et al (2012) Roles of DCL4 and DCL3b in rice phased small RNA biogenesis. Plant J 69:462–474
5. Song X, Wang D, Ma L et al (2012) Rice RNA-dependent RNA polymerase 6 acts in small RNA biogenesis and spikelet development. Plant J 71:378–389
6. Kakrana A, Li P, Patel P et al (2017). PHASIS: a computational suite for de novo discovery and characterization of phased, siRNA-generating loci and their miRNA triggers. bioRxiv https://doi.org/10.1101/158832
7. Bolger AM, Lohse M, Usadel B (2014) Trimmomatic: a flexible trimmer for Illumina sequence data. Bioinformatics 30:2114–2120
8. Axtell MJ (2013) ShortStack: comprehensive annotation and quantification of small RNA genes. RNA 19:740–751
9. Li H, Durbin R (2010) Fast and accurate long-read alignment with burrows-wheeler transform. Bioinformatics 26:589–595
10. Danecek P, Bonfield JK, Liddle J et al (2021) Twelve years of SAMtools and BCFtools. Gigascience 10:giab008
11. Quinlan AR, Hall IM (2010) BEDTools: a flexible suite of utilities for comparing genomic features. Bioinformatics 26:841–842
12. Chen C, Ridzon DA, Broomer AJ et al (2005) Real-time quantification of microRNAs by stem-loop RT-PCR. Nucleic Acids Res 33: e179
13. Varkonyi-Gasic E, Wu R, Wood M et al (2007) Protocol: a highly sensitive RT-PCR method for detection and quantification of microRNAs. Plant Methods 3:12
14. Kim WS, Haj-Ahmod Y (2014). Evaluation of plant RNA integrity number (RIN) generated using an Agilent BioAnalyzer 2100. Norgen Biotek Corp Application Note 80

Chapter 15

Expression and Purification of pAG-MNase for CUT&RUN

Taiji Kawakatsu

Abstract

Cleavage Under Targets and Release Using Nuclease (CUT&RUN) is a chromatin profiling strategy that releases specific DNA–protein complexes from in situ chromatin to the supernatant. CUT&RUN employs the hybrid protein pAG-MNase, composed of IgG binding proteins A and G, followed by micrococcal nuclease. Here, I describe the expression and purification of recombinant pAG-MNase as well as the assessment of its enzymatic activity. Purified pAG-MNase is hyperactive and enables large-scale analyses.

Key words CUT&RUN, Epigenome, Histone modification, pAG-MNase, Recombinant protein purification

1 Introduction

The epigenome, crucial for gene expression regulation, plays a pivotal role in various biological processes, such as development and stress response. Cleavage Under Targets and Release Using Nuclease (CUT&RUN) and Cleavage Under Targets and Tagmentation (CUT&Tag) provide rapid alternatives to traditional chromatin immunoprecipitation (ChIP) [1–3]. In ChIP, chromatin is solubilized by fragmentation, and chromatin complex including target proteins is immunoprecipitated [4]. Unlike ChIP, CUT&RUN directly cleaves and releases the chromatin complex containing target proteins from in situ chromatin [1]. CUT&Tag is similar to CUT&RUN, but CUT&Tag cleaves and tags the DNA around target proteins in situ [3]. The key reagent of CUT&RUN is pAG-MNase, which is the hybrid protein composed of protein A, protein G, and micrococcal nuclease (MNase) [2]. In CUT&RUN, the antibody and pAG-MNase diffuse in permeabilized nuclei, and the antibody binds to target proteins in nuclei. The pAG-MNase binds to the antibody via protein A and/or protein G and is specifically tethered to chromatin where target proteins bind to. Upon Ca^{2+} activation, pAG-MNase cleaves chromatin, releasing

Taiji Kawakatsu (ed.), *Rice: Methods and Protocols*, Methods in Molecular Biology, vol. 2869,
https://doi.org/10.1007/978-1-0716-4204-7_15,

the DNA–target protein complex from nuclei. Here I describe the method to express pAG-MNase in *E. coli* and to purify recombinant pAG-MNase. Purification of recombinant pAG-MNase involves three steps: cell lysis, affinity tag purification, and buffer exchange. This method replaces sonication with detergent for cell lysis. Additionally, this method utilizes a pre-packed desalting column for buffer exchange instead of overnight dialysis. Importantly, cell lysis by detergent has no negative effect on enzymatic activity compared to cell lysis by sonication. The amount of purified pAG-MNase by this method is enough for over 50,000 CUT&RUN assays, enabling high-throughput chromatin profiling [5].

2 Materials

2.1 Expression and Purification of pAG-MNase

1. pAG/MNase expression vector (Addgene #123461).
2. T7 Express *lysY/I^q^* Competent *E. coli* (New England Biolabs).
3. LB solid medium containing 50 μg/mL kanamycin (*see* **Note 1**).
4. LB liquid medium containing 50 μg/mL kanamycin (*see* **Note 1**).
5. Cell culture tube: Dry-heat sterilize or autoclave.
6. 1 L Erlenmeyer flask: Dry-heat sterilize or autoclave.
7. 1 M isopropyl β-D-thiogalactopyranoside (IPTG).
8. Lysis buffer: 10 mM Tris–HCl, pH 7.5, 300 mM NaCl, 10 mM imidazole, 0.14 M 2-mercaptoethanol, 1× complete protease inhibitor (Roche). Add 2-mercaptoethanol and complete protease inhibitor just before use.
9. Wash buffer: 10 mM Tris–HCl, pH 7.5, 300 mM NaCl, 20 mM imidazole, 1× complete protease inhibitor. Add complete protease inhibitor just before use.
10. Elution buffer: 10 mM Tris–HCl, pH 7.5, 300 mM NaCl, 250 mM imidazole, 1× complete protease inhibitor. Add complete protease inhibitor just before use.
11. Final buffer: 10 mM Tris–HCl, pH 7.5, 150 mM NaCl, 1 M EDTA, 1 mM phenylmethylsulfonyl fluoride (PMSF). Add PMSF just before use.
12. CelLytic B cell lysis reagent (10×) (Sigma C8740) (*see* **Note 2**).
13. His GraviTrap column (Cytiva).
14. PD-10 desalting column (Cytiva).
15. Glycerol.

2.2 Purity Check and Quantification of pAG-MNase

1. Power supply.
2. Protein gel electrophoresis cell.
3. 12% SDS-PAGE gel (*see* **Note 3**).
4. Electrophoresis buffer: 25 mM Tris, 192 mM glycine, 0.1% (w/v) SDS.
5. Protein ladder marker.
6. 2× Laemmli sample buffer: 120 mM Tris–HCl, pH 6.8, 4% (w/v) SDS, 20% glycerol, 0.02% (w/v) bromophenol blue.
7. Protein standard: Dilute 10% BSA in sterile distilled water (10 mg/mL) to 0.025, 0.125, 0.25, 0.5, 0.75, and 1 mg/mL with sterile distilled water. Add equal volume of 2× Laemmli sample buffer. Boil or incubate at 95 °C for 5 min.
8. CBB stain kit.

2.3 Assessment of pAG-MNase Activity

1. Power supply.
2. Agarose gel electrophoresis system.
3. Genomic DNA (300 ng/μL) (*see* **Note 4**).
4. DNA ladder marker.
5. MNase (New England Biolabs).
6. Tris-acetate EDTA (TAE) buffer: 40 mM Tris, 20 mM acetate, 1 mM EDTA.
7. Agarose gel: 1% agarose in TAE buffer containing 0.5 μg/mL ethidium bromide (*see* **Note 5**).

3 Methods

3.1 Expression of pAG-MNase

1. Transform pAG/MNase expression vector to T7 Express *lysY/I^q* Competent *E. coli* following the manufacturer's protocol (*see* **Note 6**).
2. Pick a single colony and inoculate a culture with 2 mL of LB liquid medium containing 50 μg/mL kanamycin in a cell culture tube.
3. Incubate at 37 °C with shaking overnight.
4. Add 2 mL of overnight culture to 200 mL of LB liquid medium containing 50 μg/mL kanamycin in a 1 L Erlenmeyer flask.
5. Incubate at 37 °C with shaking at 200 rpm for 2–3 h until OD600 reaches ~0.6.
6. Add 400 μL of 1 M IPTG (*see* **Note 7**).
7. Incubate at 37 °C with shaking at 200 rpm for 2 h.
8. Centrifuge at 5000× *g* at 4 °C for 10 min (*see* **Note 8**).

9. Discard the supernatant.
10. Proceed to Subheading 3.2 or snap-freeze in liquid nitrogen and store at −80 °C.

3.2 His-Tag Purification of pAG-MNase

1. Resuspend the pellet in 5 mL of lysis buffer by vortexing or pipetting (*see* **Note 8**).
2. Add 500 μL of CelLytic B cell lysis reagent.
3. Rotate or gently shake on a seesaw shaker at room temperature for 10 min (*see* **Note 9**).
4. Centrifuge at 12,000× *g* at 4 °C for 10 min.
5. Collect the supernatant. Take 10 μL for extract efficiency check.
6. Repeat **steps 1–4** (*see* **Note 9**).
7. Pool the supernatants from **steps 5** and **6**.
8. Suspend the pellet in 5 mL of lysis buffer by vortexing or pipetting (*see* **Note 10**). Take 10 μL for extract efficiency check.

   Hereafter, perform at 4–8 °C.
9. Cut off the tip of a His GraviTrap column, remove the cap, and discard the storage buffer by gravity flow.
10. Add 10 mL of ice-cold lysis buffer to the column and allow the buffer to flow out for equilibration.
11. Add the sample from **step** 7 to the column for binding of His-tagged pAG-MNase to Nickel Sepharose resin.
12. Add 10 mL of wash buffer to the column and allow the buffer to flow out.
13. Repeat **step 12**.
14. Add 3 mL of elution buffer to the column and collect the flowthrough. This flowthrough contains purified pAG-MNase. Take 10 μL for purity check.

3.3 Purity Check

1. Add 10 μL of 2× Laemmli sample buffer to protein samples from **steps 5**, **6**, **8**, and **14** in Subheading 3.2.
2. Boil or incubate at 95 °C for 5 min (*see* **Note 11**).
3. Assemble the SDS-PAGE electrophoresis system.
4. Load 10 μL of protein samples from **step 1** into each well of a 12% SDS-PAGE gel.
5. Run electrophoresis at 200 V for 50 min.
6. Disassemble the system and remove the gel.
7. Stain the gel with CBB stain kit following the manufacturer's protocol.
8. Check the purity of purified pAG-MNase (*see* Fig. 1).

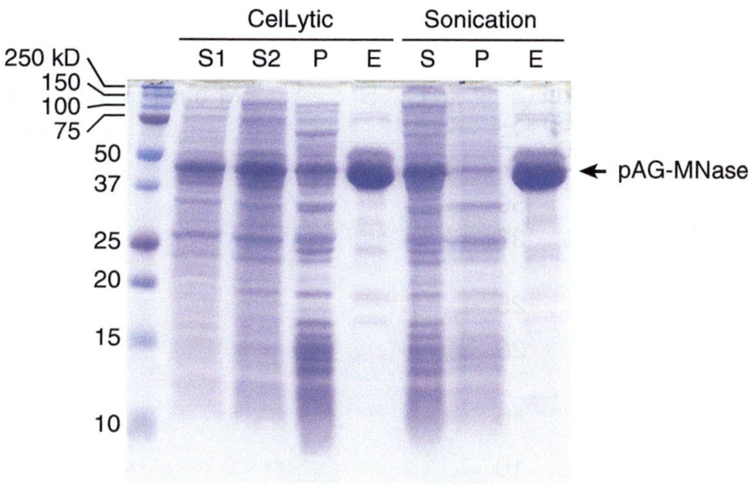

Fig. 1 Purity check of pAG-MNase by SDS-PAGE. S1, S2, and S: the supernatant containing soluble fraction after the first (S1) and the second (S2) cell lysis by CelLytic and by sonication (S). P: the pellet containing insoluble fraction after cell lysis. E: the eluted fraction. The bands corresponding to pAG-MNase with the estimated molecular weight of 44.9 kD (arrow) dominate the elution fraction, indicating high purity. The second CelLytic cell lysis effectively extracted the soluble proteins including pAG-MNase

3.4 Removal of Imidazole

1. Cut off the tip of two PD-10 desalting columns, remove the cap, and discard the storage buffer by gravity flow (*see* **Note 12**).
2. Add final buffer to fill the PD-10 desalting column and allow the buffer to flow out for equilibration (*see* **Note 13**).
3. Repeat **step 2** four times (*see* **Note 13**).
4. Add 2 mL of final buffer to purified pAG-MNase and mix by inverting (*see* **Note 14**).
5. Add 2.5 mL of purified pAG-MNase from **step 14** in Subheading 3.2 to each column and allow the buffer to flow out.
6. Add 3.5 mL of final buffer to each column and collect the flowthrough.
7. Pool the flowthroughs, add 7 mL of glycerol, and mix by rotation at 4 °C (*see* **Note 15**).
8. Prepare 100 μL aliquots and store −20 °C or − 80 °C (*see* **Note 16**).

3.5 Quantification of pAG-MNase

1. Mix 10 μL of pAG-MNase from **step 7** in Subheading 3.4 and 10 μL of 2× Laemmli sample buffer (*see* **Note 17**).
2. Mix 1 μL of MNase, 9 μL of sterile distilled water, and 10 μL of 2× Laemmli sample buffer (*see* **Note 17**).

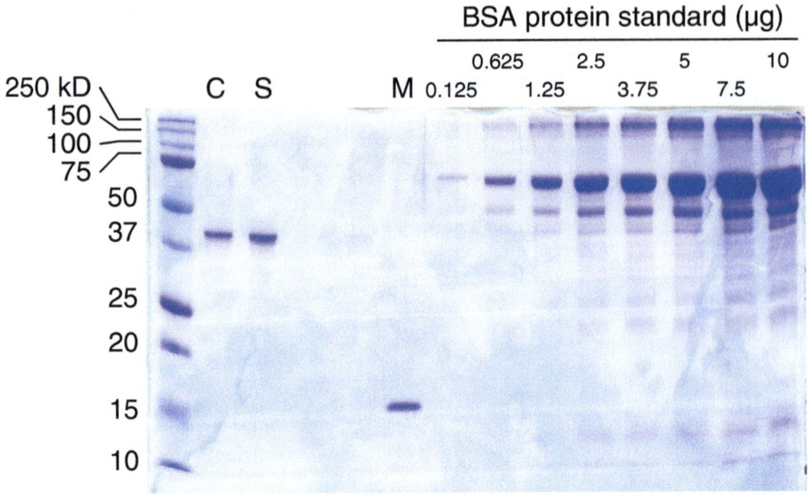

Fig. 2 Quantification of purified pAG-MNase by SDS-PAGE. C and S: pAG-MNase extracted by CelLytic (C) and sonication (S). M: MNase purchased from NEB. The quantity of diluted BSA protein standards is indicated

3. Boil or incubate protein samples from **steps 1** and **2** at 95 °C for 5 min (*see* **Note 11**).
4. Assemble the SDS-PAGE electrophoresis system (*see* Chapter 5).
5. Load 2 μL of the pAG-MNase sample from **step 3** and 10 μL of the MNase sample from **step 3** and protein standards into each well of a 12% SDS-PAGE gel (*see* **Note 17**).
6. Run electrophoresis at 200 V for 50 min.
7. Disassemble the system and remove the gel.
8. Stain the gel with CBB stain kit following the manufacturer's protocol.
9. Estimate the concentration of purified pAG-MNase (*see* Fig. 2 and **Note 18**).

3.6 Assessment of MNase Activity

This section compares the enzymatic activity of commercially available MNase and purified pAG-MNase by serial dilution (*see* Fig. 3).

1. Aliquot 10 μL of buffer EB to eight PCR tubes.
2. Add 5 μL of MNase or pAG-MNase to the first tube and mix well by pipetting.
3. Transfer 5 μL of the enzyme solution to the next tube and mix by pipetting.
4. Repeat **step 3** six times. The 1:3 dilution series of MNase and pAG-MNase is prepared in **steps 2–4**.

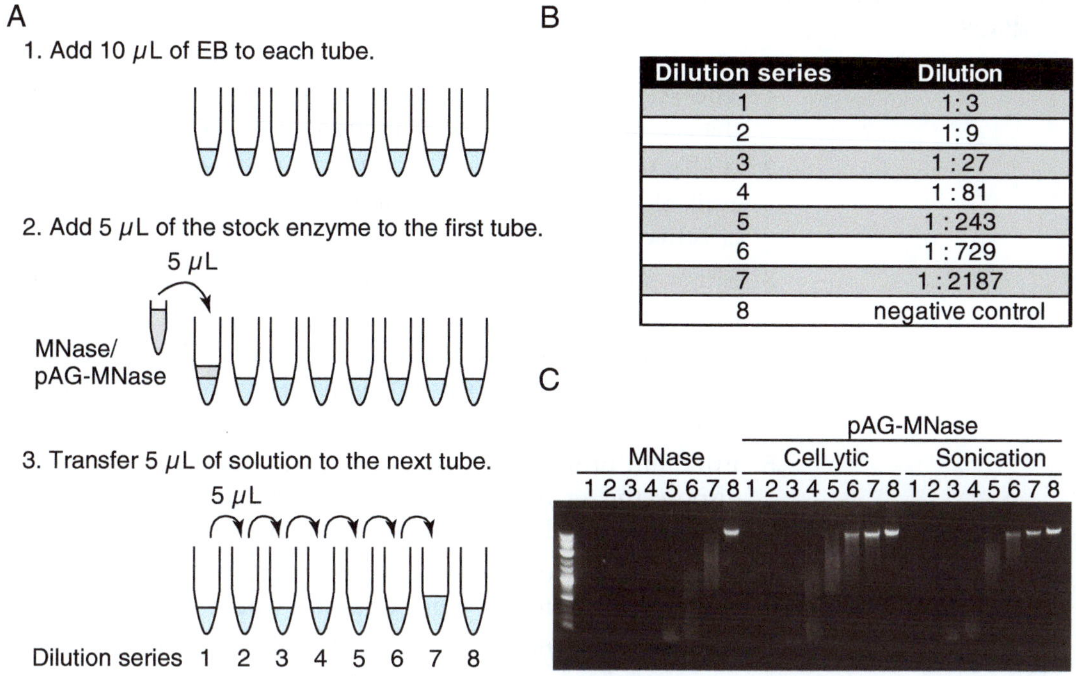

Fig. 3 Nuclease activity of purified pAG-MNase. (**a**) The scheme for preparation of the serial dilution of enzyme. First, add 10 μL of EB to eight tubes. Second, add 5 μL of the stock enzyme to the first tube and mix well by pipetting. Third, transfer 5 μL of the solution to the next tube and mix well by pipetting to dilute the solution at the ratio of 1:3. Repeat this dilution process six times. (**b**) Dilution ratio relative to the concentration of the stock enzyme. (**c**) The size distribution of fragmented rice genomic DNA by nuclease activity

5. Prepare MNase reaction master mix per reaction as follows:

10× MNase buffer	2 μL
Genomic DNA (300 ng/μL)	1 μL
Nuclease-free water	17 μL
Total	20 μL

6. Aliquot 20 μL of MNase reaction master mix to new eight PCR tubes on ice.
7. Add 1 μL of MNase or pAG-MNase dilution series and mix by pipetting.
8. Incubate at 37 °C for 15 min.
9. Add 1 μL of 0.5 M EGTA and mix by pipetting.
10. Incubate at 65 °C for 10 min.
11. Run agarose gel electrophoresis for 20 min (*see* Fig. 3 and **Note 19**).

4 Notes

1. The selection marker for pAG-MNase is kanamycin, rather than ampicillin that is widely used for protein expression vector.
2. While I have not tested, it is likely that other commercially available detergent-based cell lysis reagents would also be effective.
3. Either purchase a pre-cast gel or prepare 12% SDS-PAGE gels (*see* Chapter 5).
4. I extracted genomic DNA from rice leaves using the DNeasy Plant Mini kit (Qiagen), but high-quality commercial DNA, such as lambda DNA, is also suitable.
5. Ethidium bromide is potentially toxic. Handle with care and dispose the agarose gel and TAE buffer containing ethidium bromide, according to institutional regulations.
6. While I have not tested, it is likely that other commercially available *E. coli* for protein expression would be fine.
7. Final concentration is set at 2 mM. Although using a higher concentration may induce the reaction more efficiently, it could also increase the risk of insolubility. If other *E. coli* strain is used, optimization of IPTG concentration is required.
8. If a large-volume centrifuge is not available, split the 200 mL of cell culture into four 50 mL falcon tubes for centrifugation. Afterward, resuspend the pellet in 1.25 mL of lysis buffer and combine them into a single tube.
9. Although I do not include lysozyme treatment, it could potentially improve cell lysis. After incubation, the solution should appear slightly transparent and less viscous, indicating successful cell lysis.
10. The pellet after cell lysis should be significantly smaller than the one in **step 9** of Subheading 3.1. This pellet contains the insoluble fraction and is hard to resuspend well.
11. Be careful when handling tubes after boiling, as the cap may open suddenly due to pressure.
12. Dialysis is effective but time-consuming and requires a large buffer volume. Desalting column is quicker, taking 1–2 h, and requires less buffer. I have found the desalting is enough for subsequent CUT&RUN.
13. The LabMate PD-10 Buffer Reservoir facilitates adding 25 mL of final buffer for equilibration at once.
14. The sample volume for PD-10 column is 2.5 mL, but elution volume from the His GraviTrap is 3 mL. To load as much as the sample, mess up to 5 mL and load to two PD-10 columns.

15. Glycerol is highly vicious. Add glycerol by decant.
16. For a long-term storage, −80 °C is recommended. However, pAG-MNase is stable at −20 °C at least for 1 year.
17. The pAG-MNase is diluted to a ratio of 1:2, whereas MNase is diluted to a ratio of 1:20. Therefore, 1 μL of the pAG-MNase stock and 0.5 μL of the MNase stock are loaded to the SDS-PAGE.
18. The estimated concentrations are approximately 0.5 μg/μL for pAG-MNase (CelLytic), 0.6 μg/μL for pAG-MNase (sonication), and 1.3 μg/μL for MNase.
19. The nuclease activity of MNase stock is approximately ninefold that of pAG-MNase, as indicated by the similar size distribution between MNase dilution series 5–7 and pAG-MNase dilution series 3–5.

Acknowledgments

This work was supported by JSPS KAKENHI grant numbers 23H04756, 23H04968, and 23 K23585.

References

1. Skene PJ, Henikoff S (2017) An efficient targeted nuclease strategy for high-resolution mapping of DNA binding sites. elife 6:e21856
2. Meers MP, Bryson TD, Henikoff JG et al (2019) Improved CUT&RUN chromatin profiling tools. eLife 8:e46314
3. Kaya-Okur HS, Wu SJ, Codomo CA et al (2019) CUT&tag for efficient epigenomic profiling of small samples and single cells. Nat Commun 10: 1930
4. Furey TS (2012) ChIP–seq and beyond: new and improved methodologies to detect and characterize protein–DNA interactions. Nat Rev Genet 13:840–852
5. Tonosaki K, Ono A, Nagata H (2024) Multilayered epigenetic control of persistent and stage-specific imprinted genes in rice endosperm. Nat Plants:1–15

Chapter 16

Chromatin Immunoprecipitation from Shoot Apices and Young Panicles of Rice

Katsutoshi Tsuda

Abstract

Chromatin immunoprecipitation (ChIP) assay is a standard method to examine protein–DNA interactions in vivo. Here, I describe a basic procedure for ChIP assay in rice shoot apices using an antibody against the homeodomain transcription factor *Oryza sativa* homeobox1 (OSH1). I also introduce the tips to overcome several obstacles when performing this assay for the first time and potential pitfalls that could impact the result.

Key words Chromatin immunoprecipitation (ChIP), Antibody, Immunoblotting, Shoot meristem

1 Introduction

Chromatin immunoprecipitation is a simple but powerful method to examine whether the protein of interest binds to the target loci in vivo [1, 2]. This method utilizes a reversible crosslinking between DNA binding proteins and DNA using formaldehyde followed by chromatin extraction, immunoprecipitation, de-crosslinking, and quantification of target enrichment by quantitative PCR (qPCR) or high-throughput sequencing. Once a specific antibody against the protein of interest is prepared and positive/negative control loci are identified, this method generally yields stable results, and it is easy to examine other loci. However, before fulfilling these requirements, especially when performing ChIP for the first time, it is challenging to judge whether the experiment has been successful and to determine the cause of the failure in case no enrichment is observed. In this chapter, I introduce a basic procedure of ChIP using an anti-OSH1 antibody and rice shoot meristems as an example [3] and raise several tips to overcome these obstacles. These include testing antibodies for immunoprecipitation, identifying bona fide targets, and other tips and pitfalls.

Taiji Kawakatsu (ed.), *Rice: Methods and Protocols*, Methods in Molecular Biology, vol. 2869,
https://doi.org/10.1007/978-1-0716-4204-7_16,

2 Materials

2.1 Sample Fixation

1. Sample fixation buffer: 0.4 M sucrose, 10 mM Tris–HCl, pH 8.0, 1 mM EDTA, 1% formaldehyde (*see* **Note 1**). Prepare fresh each time. Store at room temperature until use.
2. 2 M glycine. Store at −20 °C.
3. Nylon mesh.
4. Vacuum pump.

2.2 Chromatin Extraction

1. Extraction buffer 1: 0.4 M sucrose, 10 mM Tris–HCl, pH 8.0, 5 mM β-mercaptoethanol, 1× complete protease inhibitor cocktail (*see* **Note 2**).
2. Extraction buffer 2: 0.25 M sucrose, 10 mM Tris–HCl, pH 8.0, 10 mM $MgCl_2$, 1% Triton ×-100, 5 mM β-mercaptoethanol, 1× complete protease inhibitor cocktail (*see* **Note 2**).
3. Extraction buffer 3: 1.7 M sucrose, 10 mM Tris–HCl, pH 8.0, 2 mM $MgCl_2$, 0.15% Triton ×-100, 5 mM β-mercaptoethanol, 1× complete protease inhibitor cocktail (*see* **Note 2**).
4. Sodium dodecyl sulfate (SDS) nuclei lysis buffer: 50 mM Tris–HCl, pH 8.0, 1% SDS, 10 mM EDTA, 1× complete protease inhibitor cocktail (*see* **Note 2**).
5. Pestle.
6. Mortar.
7. Liquid nitrogen.
8. 50 μm cell strainer.
9. Miracloth.
10. Branson Sonifier 250 sonicator.
11. RNase A (0.2 mg/mL).
12. Agarose.
13. 0.5× TAE buffer: 20 mM Tris, 10 mM acetic acid, 25 mM EDTA.
14. Electrophoresis device.

2.3 Immunoprecipitation

1. ChIP dilution buffer: 50 mM Tris–HCl, pH 8.0, 0.167 M NaCl, 1.1% Triton ×-100, 0.11% sodium deoxycholate, 1× complete protease inhibitor cocktail (*see* **Note 2**).
2. 2× RIPA buffer: 0.1 M Tris–HCl, pH 8.0, 0.3 M NaCl, 2 mM EDTA, 0.2% SDS, 0.2% sodium deoxycholate, 2% Triton ×-100.

3. Low-salt RIPA buffer: 1× RIPA buffer, 1× complete protease inhibitor cocktail (*see* **Note 2**).
4. High-salt RIPA buffer: 1× RIPA buffer, 0.5 M NaCl, 1× complete protease inhibitor cocktail (*see* **Note 2**).
5. LNDET buffer: 0.25 M LiCl, 1% IGEPAL, 1% sodium deoxycholate, 1 mM EDTA, 10 mM Tris–HCl, pH 8.0.
6. TE buffer: 10 mM Tris–HCl, pH 8.0, 1 mM EDTA.
7. ChIP direct elution buffer: 10 mM Tris–HCl, pH 8.0, 0.3 M NaCl, 5 mM EDTA, 0.5% SDS.
8. Dynabeads Protein A (Thermo Fisher).
9. Normal rabbit IgG: This is used as a negative control.
10. Anti-OSH1 rabbit antibody.
11. 10× PBS: 1.37 M NaCl, 27 mM KCl, 100 mM Na_2HPO_4, 18 mM KH_2PO_4.
12. 1× PBS containing 0.5% (w/v) bovine serum albumin (BSA).
13. Proteinase K (20 mg/mL).

3 Methods

3.1 Sample Fixation

1. Dissect and collect approximately 30 shoot apices, each about 3 mm^3 in size, from 1-month-old plants. Each apex contains the shoot apical meristem, the two youngest leaf primordia, and young stems. Alternatively, collect five young panicles, each about 5 mm in length. Place samples into a 50 mL tube filled with water. Try to remove non-target tissues/organs (e.g., young leaves when collecting shoot meristem-enriched samples or young panicles) (*see* **Note 3**).
2. Pour into a nylon mesh bag and collect samples (*see* Fig. 1 and **Note 4**).
3. Put the bag containing samples into a 50 mL tube and add 20 mL of the sample fixation buffer. Make sure that the entire bag is covered by the sample fixation buffer. A nylon mesh bag helps to keep samples sunk.
4. Incubate for 10 min under vacuum at room temperature.
5. Add 1 mL of 2 M glycine (to a final concentration of 0.1 M) to quench fixation.
6. Incubate samples for 5 min under vacuum at room temperature.
7. Discard the sample fixation buffer.
8. Add 40 mL of distilled water, invert the tube ten times, and discard water.

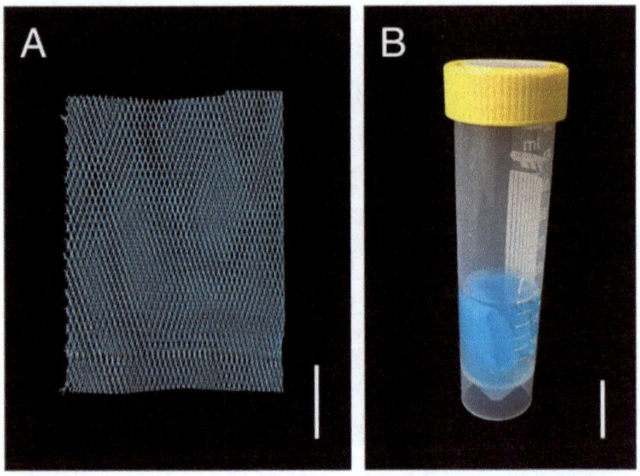

Fig. 1 A nylon mesh bag for the sample fixation. (**a**) A bag made of conventional nylon mesh for kitchen use. The edges were sealed using a heat sealer. (**b**) The mesh bag was accommodated in a 50 mL tube filled with sample fixation buffer. The bars are 2 cm

9. Repeat **step 8** twice.
10. Remove excess water by putting the sample bag on paper towels.
11. Collect the samples in a 10 mL tube (*see* **Notes 5** and **6**). Store at −80 °C.

3.2 Chromatin Extraction

1. Cool a pestle and a mortar with liquid nitrogen.
2. Grind samples into a fine powder using a pestle and a mortar (*see* **Note 7**).
3. Add 5 mL of extraction buffer 1 into the mortar and resuspend samples (*see* **Note 8**).
4. Filter the suspension through a 50 μm cell strainer placed on a 15 mL tube (*see* **Note 9**).
5. Centrifuge at 1900× *g* for 10 min at 4 °C.
6. Remove the supernatant and resuspend the pellet in 1 mL of extraction buffer 2 by pipetting up and down.
7. Transfer the suspension into a new 1.5 mL tube.
8. Centrifuge at 10,000× *g* for 10 min at 4 °C.
9. Remove the supernatant and resuspend the pellet in 300 μL of extraction buffer 2 by pipetting up and down.
10. Add 300 μL of extraction buffer 3 to a new tube and carefully overlay the sample suspension from **step 9** onto the extraction buffer 3.

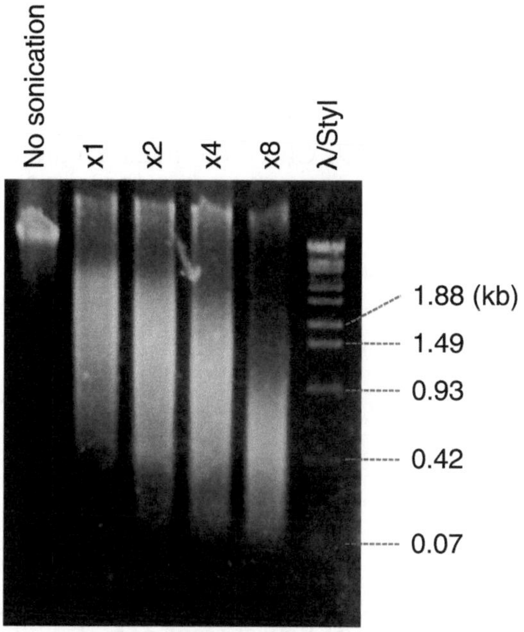

Fig. 2 Chromatin fragmentation by sonication. An electrophoresis image of chromatin samples sonicated for cycles indicated above. One cycle was for 10 s at 40% duty cycle and 20% and output level 2 in Branson Sonifier 250. The gel was stained with ethidium bromide. The size marker (λ/StyI) was indicated on the right

11. Centrifuge at 10,000× *g* for 10 min at 4 °C.
12. Remove the supernatant and resuspend the pellet in 300 μL of SDS nuclei lysis buffer by gently pipetting up and down (*see* **Note 10**). Avoid forming bubbles.
13. Incubate on ice for 20 min. Vortex samples gently every 5 min.
14. Sonicate samples using Branson Sonifier 250 (output level 2, *ON* time 10 s, duty cycle 40%, the number of cycles 4–8) to obtain chromatin fragments ranging from ~200 to 2000 bp (*see* Fig. 2 and **Note 11**). Keep samples cool by placing tubes on ice between intervals of each sonication cycle.
15. Centrifuge samples at 10,000× *g* for 10 min at 4 °C.
16. Transfer the supernatant into a new tube. This is the chromatin samples for subsequent steps.
17. Measure DNA concentration using NanoDrop. Set the blank using SDS nuclei lysis buffer.

18. Transfer 10 μL of supernatant into a new 1.5 mL tube and add 1 μL of 0.2 mg/mL RNase A. Incubate at 65 °C overnight. Store the remaining samples at −80 °C until proceeding to the immunoprecipitation steps in Subheading 3.3.
19. Check the fragmentation by agarose electrophoresis on a 1% agarose gel (*see* **Note 12**).

3.3 Immuno-precipitation

1. Resuspend Dynabeads Protein A by rotating the bottle at room temperature for 10 min.
2. Transfer 40 μL of Dynabeads Protein A to a 1.5 mL tube for each sample.
3. Add 1 mL of 1× PBS containing 0.5% BSA and mix by inverting several times (*see* **Note 13**).
4. Briefly spin down and place the tube on a magnetic rack (*see* **Note 14**).
5. After beads are completely captured and the solution is clear, discard the supernatant.
6. Repeat **steps 3–5**.
7. Repeat **steps 3–5** with 1 mL of low-salt RIPA buffer.
8. Resuspend beads with 200 μL of low-salt RIPA buffer (*see* **Note 15**).
9. Add approximately 1 μg of anti-OSH1 rabbit antibody or normal rabbit IgG (as a negative control) to the bead suspension from **step 8**.
10. Rotate for 10 min at room temperature.
11. Repeat **steps 4** and **5**.
12. Add 1 mL of low-salt RIPA buffer.
13. Resuspend by inverting tubes.
14. Repeat **steps 4** and **5**.
15. Repeat **steps 12–14** and close the lid until the diluted chromatin samples are ready.
16. Thaw chromatin samples from **step 18** in Subheading 3.2 at room temperature and vortex briefly.
17. Add 10 μg of the chromatin sample to a new 1.5 mL tube and make up to 1350 μL with chromatin dilution buffer. Mix the sample by inverting the tube and spin down.
18. Transfer 13.5 μL of the diluted chromatin (1% [w/v] of the total volume) from **step 17** into a new 1.5 mL tube as an input control.
19. Add the remaining diluted chromatin sample into the tube containing the beads bound by antibodies from **step 15**.

20. Rotate for 1 h at room temperature or for 2–3 h at 4 °C.
21. Briefly spin the tube and put it on a magnetic rack.
22. After beads are completely captured and the solution is clear, discard the supernatant.
23. Add 1 mL of low-salt RIPA buffer and rotate for 10 min at room temperature.
24. Repeat **steps 21** and **22**.
25. Repeat **steps 23** and **24** (*see* **Note 16**).
26. Add 1 mL of high-salt RIPA buffer and rotate for 10 min at room temperature.
27. Repeat **steps 21** and **22.**
28. Repeat **steps 26** and **27**.
29. Add 1 mL of LNDET buffer and rotate for 10 min at room temperature.
30. Repeat **steps 21** and **22**.
31. Repeat **steps 29** and **30**.
32. Add 1 mL of TE buffer and rotate for 10 min at room temperature.
33. Briefly spin the tube and transfer the bead suspension to a new 1.5 mL tube (*see* **Note 17**).
34. Put the tube on a magnetic rack. After beads are completely captured and the solution is clear, discard the supernatant.
35. Add 1 mL of TE buffer and rotate for 10 min at room temperature.
36. Repeat **steps 21** and **22**.
37. Add 100 μL of TE buffer to the beads. This is the ChIPed sample.
38. Add 86.5 μL of TE buffer to 13.5 μL of the diluted chromatin saved at **step 18**. This is the input control, which is 1% of the total input chromatin sample.
39. Incubate at 95 °C for 10 min.
40. Briefly spin tubes, add 1 μL of proteinase K, and mix by vortexing briefly.
41. Incubate at 43 °C for 1 h.
42. Incubate at 95 °C for 10 min to inactivate proteinase K.
43. Briefly spin tubes and put them on a magnetic rack.
44. After beads are completely captured and the solution is clear, transfer the supernatant into new tubes. Store at −20 °C.
45. Check the enrichment of the target loci relative to 1% of the total input by quantitative PCR (*see* **Notes 18–20**).

4 Notes

1. Use fresh formaldehyde solution commercially available (within a month after purchase is safe). The use of outdated formalin (e.g., a year old) results in inefficient fixation, leading to false-negative enrichment.
2. β-Mercaptoethanol and the protease inhibitor cocktail should be added just before use. Stock solutions without β-mercaptoethanol and the protease inhibitor cocktail can be stored at 4 °C for several months.
3. This will increase the exposure of your target tissues to the fixative and enrich them relative to the total sample. In case your target tissues are very small (e.g., anthers [4]), simply fix the whole material (e.g., flowers or inflorescences). In case the target tissue is embedded deeply inside (e.g., shoot apices [3]) and it takes too long to dissect the tissue, simply cutting the sample in half may work to expose it. Non-target tissues can be removed after the fixation.
4. A handmade bag made of any nylon mesh (e.g., for kitchen use) can be used (*see* Fig. 1).
5. Non-target tissues can be removed on a clean surface (e.g., a new plastic dish) under a dissecting microscope at room temperature.
6. If the sample is small and fragile, collecting samples in a tube and crushing them with stainless beads will work. This is particularly advantageous for small samples (e.g., anthers), which are prone to be lost during grinding using pestles and mortars [4]. According to Ono et al. (2018), ~1200 anthers of 0.5 mm size were collected per replicate [4]. In this case, use 2 mL screw-capped and impact-resistant tubes to avoid breakage of tubes and sample loss.
7. In case using a tube and beads, two stainless beads of ~5 mm diameter per tube usually work well to pulverize samples. Run a bead beater homogenizer at 1100 rpm for 1 to 2 min. Make sure to pre-cool the sample holder using liquid nitrogen to keep samples frozen during beads beating.
8. Add the buffer immediately after pulverization. It is anticipated that the buffer will freeze. Gently mix the sample and the buffer while melting. Adjust the volume of the buffer depending on the amount of samples (e.g., 1 mL for anthers).
9. In case the sample volume is large, filter the suspension through two layers of Miracloth placed in a funnel and collect samples in an ice-cooled 15 mL tube, first. Squeeze the Miracloth to obtain flowthrough as much as possible.

10. Samples containing many meristems (e.g., young panicles) tend to be viscous while pipetting. This is a good sign that chromatins are being extracted. In the case of other tissues, such a viscosity is not usually observed. Avoid making bubbles in samples.
11. The sonication condition depends on the sonicator. Optimize the duration and the output level of the sonicator.
12. Add 1–2 μL of glycerol instead of loading dye; otherwise the smear of fragmented chromatin will be partially masked by the dye.
13. In the case of multiple samples (genotypes of materials/antibodies/replicates), up to 400 μL of Dynabeads can be washed with 1 mL of 1× PBS containing 0.5% BSA in a single tube.
14. At each washing step, spin the tubes briefly in a minicentrifuge for 1 s to remove any liquid trapped in the lid prior to setting the tubes in the magnetic rack. This is important to minimize the carryover of buffers/samples to the next step, especially after adding chromatin samples.
15. In the case of multiple samples, multiply the volume of low-salt RIPA buffer by the number of samples. Aliquot 200 μL of the bead suspension into each tube.
16. At this point, whether the antibody works for ChIP can be assessed by immunoblotting (*see* Fig. 3). This is important in case an untested antibody is used. To do this, repeat this washing step two more times, boil the beads in 30 μL of SDS-PAGE sample buffer, and then proceed to the conventional immunoblotting (*see* Chapter 5). Mutant sample, such as null allele(s), can be a negative control. If the target protein can be detected in the immunoprecipitated (IP) fraction, it is very likely that the chromatin regions associated with this protein are also present in the IP fraction.
17. Replacing the tube reduces the background level of unbound genomic DNA.
18. If ChIP-seq is planned, purify ChIPed DNA by phenol/chloroform extraction and ethanol precipitation or by using a column-based DNA purification kit. Prepare ChIP-seq library using a low-input DNA library preparation kit, following the manufacturer's protocol. If the amount of starting materials is small, combine multiple replicates into a single batch at this step to increase the amount of "ChIPed" DNA for the library preparation.
19. If bona fide target regions of protein of interest are unknown, it is challenging to design primers. In such case, evolutionary

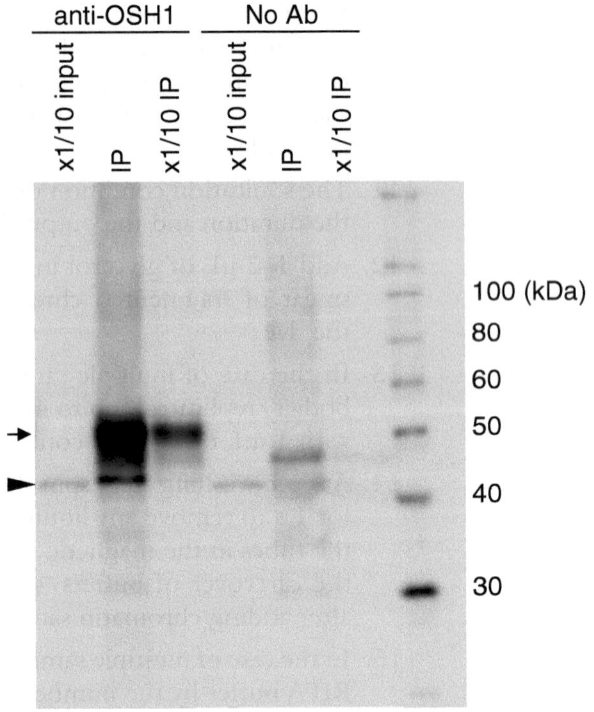

Fig. 3 Assessment of immunoprecipitation. The immunoprecipitation (IP) fraction was analyzed by immunoblotting using an anti-OSH1 antibody. An arrowhead indicates the band of OSH1 protein immunoprecipitated in the presence of the antibody. An arrow indicates the heavy chain of the rabbit IgG. The size marker was indicated on the right

conserved non-coding sequences (CNSs) around the locus of interest may be good candidates. CNSs can be identified using the VISTA browser (*see* Fig. 4a) [7].

20. Materials from null mutants or tissues that do not express the protein of your interest can be used as negative controls (*see* Fig. 4b). Including these controls is very informative to test the specificity of the antibody. Alternatively, using transgenic plants that express the GFP-tagged protein of interest is one of the promising approaches. Anti-GFP antibodies commercially available are generally very specific to GFP expressed in rice [4–6]. Manufacturers also provide information on their applicability to immunoprecipitation.

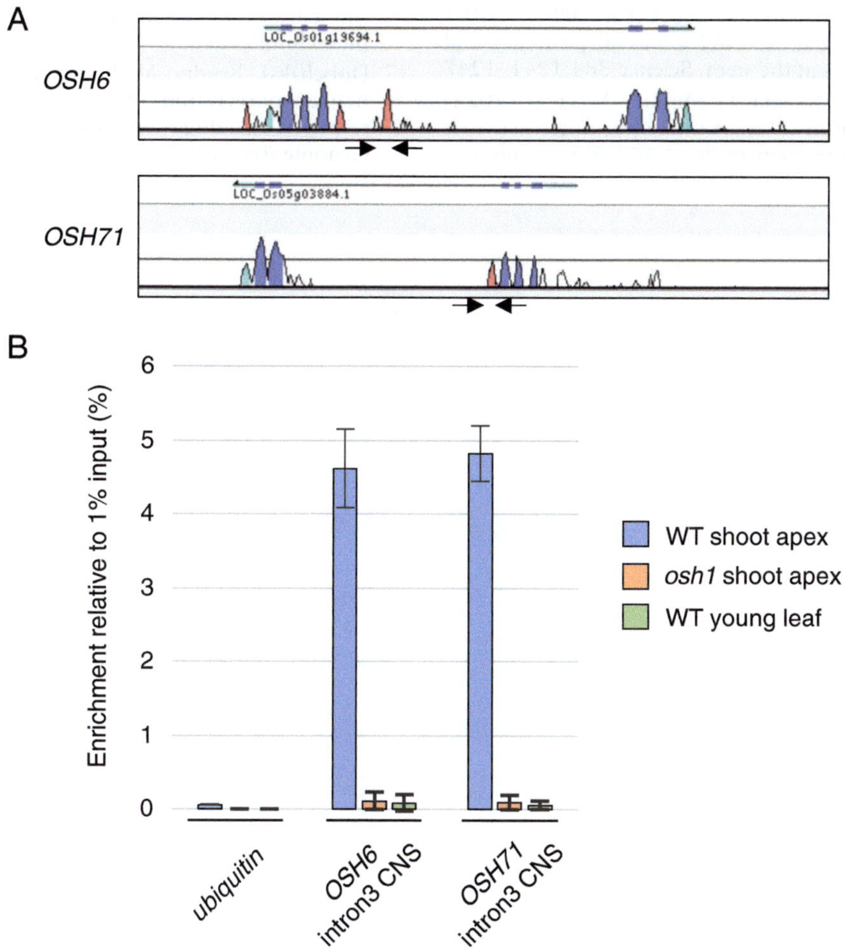

Fig. 4 Enrichment of OSH1-bound fragments in ChIPed DNA. (**a**) VISTA-Point browser views of *OSH6* and *OSH71* loci. Conserved non-coding regions above 70% of sequence conservation between *Oryza sativa* and *Brachypodium distachyon* are shown in pink below gene models. PCR primers used in (**b**) are shown with arrows. (**b**) Quantification of enrichment of target loci by qPCR. Rice *ubiquitin* locus was used as a negative control. The *osh1* mutant shoot apex and wild-type young leaves were also used as negative controls

References

1. Dahl JA, Collas P (2008) A rapid microchromatin immunoprecipitation assay (microChIP). Nat Protoc 3:1032–1045
2. Kaufmann K, Muiño JM, Østerås M et al (2010) Chromatin immunoprecipitation (ChIP) of plant transcription factors followed by sequencing (ChIP-SEQ) or hybridization to whole genome arrays (ChIP-CHIP). Nat Protoc 5: 457–472
3. Tsuda K, Ito Y, Sato Y et al (2011) Positive autoregulation of a KNOX gene is essential for shoot apical meristem maintenance in rice. Plant Cell 23:4368–4381
4. Ono S, Liu H, Tsuda K et al (2018) EAT1 transcription factor, a non-cell-autonomous regulator of pollen production, activates meiotic small RNA biogenesis in rice anther tapetum. PLoS Genet 14:e1007238

5. Tsuda K, Maeno A, Otake A et al (2024) *YABBY* and diverged *KNOX1* genes shape nodes and internodes in the stem. Science 384:1241–1247
6. Tsuda K, Suzuki T, Mimura M et al (2022) Comparison of constitutive promoter activities and development of maize ubiquitin promoter- and gateway-based binary vectors for rice. Plant Biotechnol (Tokyo) 39:139–146
7. Dubchak I, Brudno M, Loots GG et al (2000) Active conservation of noncoding sequences revealed by three-way species comparisons. Genome Res 10:1304–1306

Chapter 17

In Vitro Genome-Wide Identification of Transcription Factor Binding Sites

Taiji Kawakatsu

Abstract

DNA affinity purification followed by sequencing (DAP-seq) profiles transcription factor (TF) binding sites in vitro, in which recombinant TF captures the genomic fragments with TF binding sites. Here, I describe the preparation of pre-PCR DNA-seq libraries for 96 TFs, including the expression and purification of recombinant TF, the affinity purification of TF binding fragments, and the sequencing-ready library preparation. Using this method, up to 96 DAP-seq libraries can be prepared in 2 days.

Key words DAP-seq, Transcription factor, Gene regulatory network, Cell-free protein expression, Halo-tag

1 Introduction

A gene regulatory network (GRN) is a complex network among a collection of genes and their products that interact and regulate each other [1, 2]. GRNs fundamentally underlie biological processes, such as development, response to stimuli, and disease tolerance. Transcription factors (TFs) are proteins that bind to specific DNA sequences to activate or repress gene expression [3]. Since TFs regulate the expression of other genes as well as their own, TFs are key players that shape the circuitry composed of multistep cascades in GRNs [3, 4]. Therefore, globally identifying target genes of TFs is essential for the understanding of the comprehensive circuitries in an organism.

Chromatin immunoprecipitation followed by sequencing (ChIP-seq) has been the gold standard to study the protein–DNA interactions. ChIP-seq experiments include crosslinking of TF–DNA complex, nuclei isolation, chromatin fragmentation, and immunoprecipitation [5]. ChIP-seq requires a specific antibody against TF. However, preparing antibodies or generating

Taiji Kawakatsu (ed.), *Rice: Methods and Protocols*, Methods in Molecular Biology, vol. 2869,
https://doi.org/10.1007/978-1-0716-4204-7_17,

transgenic plants expressing the tag-fused TFs for many TFs is unfeasible.

To overcome this limitation, DNA affinity purification followed by sequencing (DAP-seq) has been developed [6, 7]. DAP-seq employs in vitro expression of Halo-tag-fused TFs eliminating the requirement of specific antibodies against TFs and generation of transgenic plants. The highly stable covalent bond formation between the Halo-ligand and the Halo-tag under native condition simplifies the purification of DNA fragments compared to ChIP-seq. The success rate of DAP-seq highly depends on TF family (high in bZIP and NAC, low in bHLH and MADS), and the overall success rate is around 40% [6]. Although DAP-seq is an in vitro assay and does not reflect in vivo specific effects, such as interaction between other TFs as well as epigenomes other than DNA methylation, its consistency with ChIP-seq has been experimentally validated [6, 7]. Furthermore, DAP-seq captures the direct TF–DNA interaction, rather than the TF complex–DNA interaction. Recently, DAP-seq has also been utilized to study TF heterodimer–DNA interactions and TF–methylated DNA interactions [8–10].

Here, I describe the DAP-seq library preparation method. This method includes preparation of pre-PCR DNA-seq library as an input for DAP-seq, in vitro expression of Halo-tagged TFs, Halo-tag purification of TFs, DNA affinity purification with Halo-tagged TFs, library amplification, and optional verification of TF expression. Using this method in a 96-well PCR plate format, DAP-seq libraries of 96 TFs can be prepared in 2 days. This method can be utilized for small-scale experiments on a few TFs as well as large-scale projects on hundreds of TFs [6, 11–15].

2 Materials

2.1 Equipment

1. Covaris Focused-ultrasonicator M220 (Covaris).
2. M220 Holder XTU (Covaris).
3. M220 Holder XTU Insert microTUBE 130 μL (Covaris).
4. Thermal cycler.
5. Plate rotator.
6. DynaMag-2 magnet stand (Invitrogen).
7. DynaMag-96 Side magnet stand (Invitrogen).
8. Multichannel pipette: 8 or 12 channels, capacity 0.5–10 μL, 1–20 μL, 5–50 μL, 30–300 μL.
9. Qubit Fluorometer (Invitrogen).
10. Power supply (optional; *see* **Note 1**).
11. Mini-PROTEAN Tetra Cell (Bio-Rad) (optional; *see* **Note 1**).

12. Trans-Blot Turbo system (Bio-Rad) (optional; *see* **Note 1**).
13. LAS 4000 (Cytiva) (optional; *see* **Note 1**).

2.2 Consumables

1. microTUBE AFA Fiber Pre-Slit Snap-Cap (Covaris).
2. 1.5 mL DNA LoBind tube (Eppendorf).
3. 0.2 mL PCR tube.
4. PCR 8-tube strip.
5. 96-well PCR plate, no-skirt, thin-walled.
6. Adhesive PCR plate seal.
7. Reagent reservoir, capacity 50 mL.

2.3 Preparation of Pre-PCR DNA-Seq Library

1. SPRI beads: Prepare as described in **item** 7 in Subheading 2.3 of Chapter 2. SPRI beads are used for nucleic acid purification after any enzymatic reaction.
2. 80% ethanol.
3. Elution buffer (EB): 10 mM Tris–HCl, pH 8.5.
4. Rice genomic DNA.
5. NEBNext Ultra II DNA Library Prep Kit for Illumina (New England Biolabs) (*see* **Note 2**).
6. NEBNext Multiplex Oligos for Illumina (New England Biolabs) (*see* **Note 3**).
7. Qubit dsDNA High Sensitivity Assay Kit (Invitrogen).

2.4 Expression of Recombinant TF

1. TNT SP6 High-Yield Wheat Germ Protein Expression System (Promega).
2. Recombinant TF expression vector: The pIX-Halo vector, which includes the TF coding sequence. Transfer the TF coding sequence from the Gateway Entry vector to the pIX-Halo-ccdB-CmR vector (ABRC CD3-1742) using Gateway LR Clonase II (Invitrogen), according to the manufacturer's protocol. Ensure that the TF coding sequence is in-frame and N-terminally fused with the Halo-tag. Extract plasmid DNAs using a column-based kit. Alternatively, amplify DNA fragments containing SP6 promoter, Halo-tagged TF coding sequence, and poly A sequence using a high-fidelity PCR enzyme and T7 promoter oligo (5′-taatacgactcactatagg-3′) and T7 terminator reverse oligo (5′-caaaaaacccctcaagacccg-3′), according to the manufacturer's protocol. Dilute to 100–200 ng/μL with EB.

2.5 Halo-Tag Purification and DNA Affinity Purification

1. Magne HaloTag Beads (Promega).
2. PBS: 137 mM NaCl, 8.1 mM Na_2HPO_4, 2.7 mM KCl, 1.5 mM KH_2PO_4, pH 7.4.

3. PBS-X: PBS containing 0.005% Triton X-100. Store at room temperature up to 1 week.
4. Elution buffer (EB): 10 mM Tris–HCl, pH 8.5.

2.6 DAP-seq Library Amplification

1. KAPA HiFi HotStart ReadyMix (Roche).
2. NEBNext Multiplex Oligos for Illumina (New England Biolabs) (*see* **Note 3**).
3. SPRI beads: Prepare as described in **item** 7 in Subheading 2.3 of Chapter 2. SPRI beads are used for nucleic acid purification after any enzymatic reaction.
4. 80% ethanol.
5. Elution buffer (EB): 10 mM Tris–HCl, pH 8.5.
6. Qubit dsDNA High Sensitivity Assay Kit (Invitrogen).

2.7 (Optional) Verification of TF Expression (See Note 4)

1. 4–15% Mini-PROTEAN TGX Precast Gel (Bio-Rad).
2. Electrophoresis buffer: 25 mM Tris, 192 mM glycine, 0.1% (w/v) SDS.
3. Trans-Blot Turbo Mini 0.2 μm PVDF Transfer Packs (Bio-Rad).
4. TBS-T: 25 mM Tris–HCl, pH 7.4, 137 mM NaCl, 2.68 mM KCl, 0.05% (v/v) Tween 20.
5. Blocking solution: 5% (w/v) skimmed milk in TBS-T.
6. Primary antibody: Anti-Halo rabbit antibody (Promega).
7. Secondary antibody: Anti-rabbit IgG, HRP-linked (Cell Signaling Technology).
8. Clarity Western ECL Substrate (Bio-Rad).

3 Methods

The outline of this method is depicted in Fig. 1. Ensure that the pre-PCR DNA-seq library is prepared as described in Subheading 3.1 at least 1 day before proceeding to the subsequent sections.

3.1 Preparation of Pre-PCR DNA-seq Library

1. Turn on the Covaris Focused-Ultrasonicator M220.
2. Set the M220 Holder XTU and the M220 Holder XTU Insert microTUBE 130 μL.
3. Fill the water tank with distilled water.
4. Select the 200 bp target mode and wait until the water temperature reaches 20 °C (*see* **Note 5**).
5. Transfer 5 μg of genomic DNA in 130 μL of EB to two microTUBEs AFA Fiber Pre-Slit Snap-Cap (*see* **Note 6**).

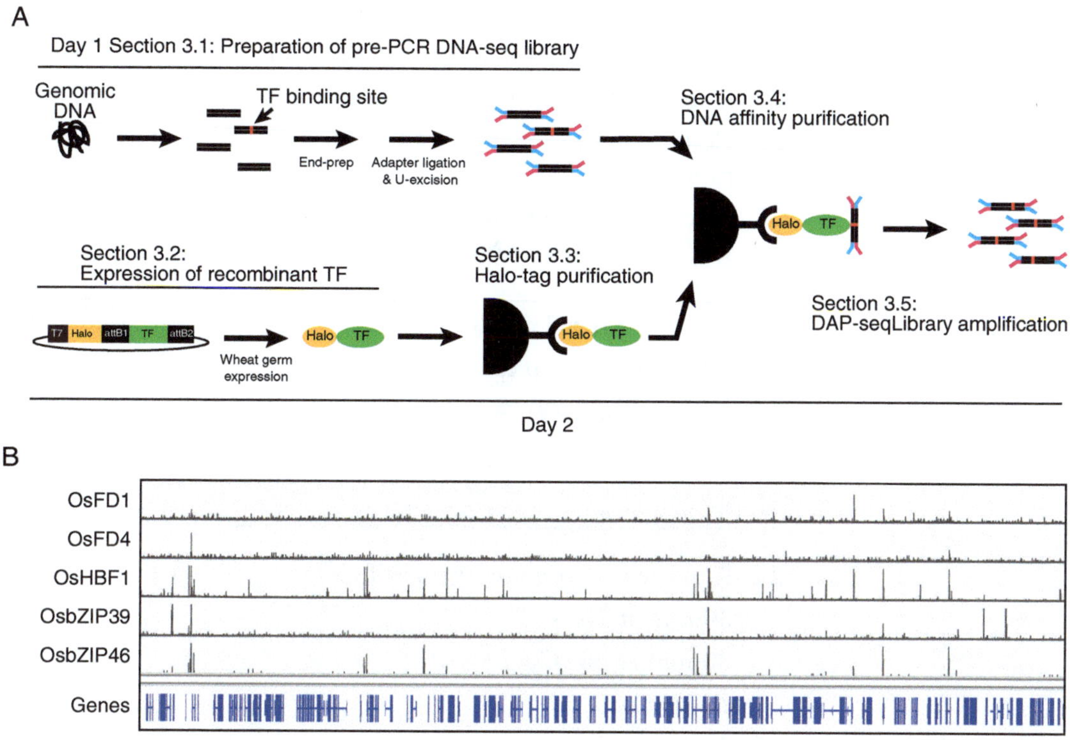

Fig. 1 The outline of DAP-seq library preparation. (**a**) Perform pre-PCR DNA-seq library preparation (Subheading 3.1) in day 1 and other steps (Subheadings 3.1, 3.2, 3.3, 3.4, and 3.5) in day 2. (**b**) The example browser snapshot of DAP-seq results for bZIP TFs. (Data are obtained from Refs. [13, 14])

6. Run the Covaris to sonicate genomic DNA.
7. Combine the contents of the two microTUBEs containing the sonicated genomic DNA into a single 1.5 mL DNA LoBind tube.
8. Add 470 μL of SPRI beads (approximately 1.8× volume).
9. Rotate at room temperature for 10 min (*see* **Note 7**).
10. Briefly spin down and place the tube on DynaMag-2 for 5 min.
11. Discard the supernatant.
12. Add 1 mL of 80% ethanol to the tube and wait 30 s.
13. Discard the supernatant.
14. Repeat **steps 12** and **13**.
15. Briefly spin down and place the tube on DynaMag-2.
16. Remove the residual 80% ethanol as much as possible (*see* **Note 8**).
17. Keep the lid open and air-dry at room temperature for 5 min or until the bead pellet dries (*see* **Note 9**).
18. Remove the tube from DynaMag-2.

19. Add 165 μL of elution buffer and mix well by pipetting.
20. Incubate at room temperature for 2 min.
21. Place the tube on DynaMag-2 magnet for 2 min.
22. Collect 163 μL of the supernatant to a 1.5 mL DNA LoBind tube.
23. Assemble the end-prep reactions in each tube of a PCR 8-tube strip as follows (*see* **Note 10**):

NEBNext ultra II end prep reaction buffer	7 μL
NEBNext ultra II end prep enzyme mix	3 μL
Fragmented genomic DNA	20 μL
Nuclease-free distilled water	30 μL
Total	60 μL

24. Run the end-prep reaction as follows:

    30 min at 25 °C

    30 min at 65 °C

    Hold at 4 °C
25. Add the following components to each tube from **step 23**:

End-prepped genomic DNA from **step 24**	60 μL
NEBNext adapter for Illumina	2.5 μL
NEBNext ultra II ligation master mix	30 μL
NEBNext ligation enhancer	1 μL
Total	93.5 μL

26. Run the adapter ligation reaction as follows:

    15 min at 25 °C

    Hold at 4 °C
27. Add 3 μL of USER enzyme to the adapter ligation reaction from **step 26** (*see* **Note 11**).
28. Mix well by pipetting and incubate at 37 °C for 15 min.
29. Add 75 μL of SPRI beads and mix well by pipetting.
30. Incubate at room temperature for 10 min.
31. Briefly spin down and place the tube on DynaMag-96 for 5 min.
32. Discard the supernatant.
33. Add 200 μL of 80% ethanol and wait 30 s.
34. Discard the supernatant.

35. Repeat **steps 31** and **32**.
36. Briefly spin down and place the tube on DynaMag-96.
37. Remove the residual 80% ethanol as much as possible.
38. Keep the lid open and air-dry at room temperature for 3 min or until the bead pellet dries.
39. Remove the tubes from DynaMag-96.
40. Add 51 μL of EB and mix well by pipetting.
41. Incubate at room temperature for 2 min.
42. Place the tubes on DynaMag-96 for 2 min.
43. Collect 50 μL of the supernatant and combine them to a single 1.5 mL DNA LoBind tube.
44. Measure the concentration with a Qubit Fluorometer and Qubit dsDNA High Sensitivity Assay Kit. Store the remaining pre-PCR DNA-seq library at −20 °C.

3.2 Expression of Recombinant TF

1. Thaw TNT SP6 High-Yield Wheat Germ Master Mix at room temperature (*see* **Note 12**).
2. Assemble the in vitro expression reaction as follows:

TNT SP6 high-yield wheat germ master mix	15 μL
Recombinant TF expression vector (100–200 ng/μL)	10 μL
Total	25 μL

3. Incubate at 25 °C for 2.5 h (*see* **Note 13**).

3.3 Halo-Tag Purification

1. Resuspend 1 mL of Magne HaloTag Beads and place on DynaMag-2 for 30 s.
2. Discard the supernatant.
3. Add 1 mL of PBS-X.
4. Remove the tube from DynaMag-2 and mix well by pipetting.
5. Repeat **steps 3** and **4** twice.
6. Add 850 μL of PBS-X and aliquot 130 μL to each tube of a PCR 8-tube strip (*see* **Note 14**).
7. Dispense 10 μL of washed Magne HaloTag Beads to the bottom of each well of a 96-well PCR plate using a multichannel pipette (*see* **Note 15**).
8. Dispense 50 μL of PBS-X to each well using a multichannel pipette (*see* **Note 15**).
9. Transfer 20 μL of the in vitro expression reaction from **step 3** in Subheading 3.2.
10. Seal the plate with adhesive PCR plate seal.

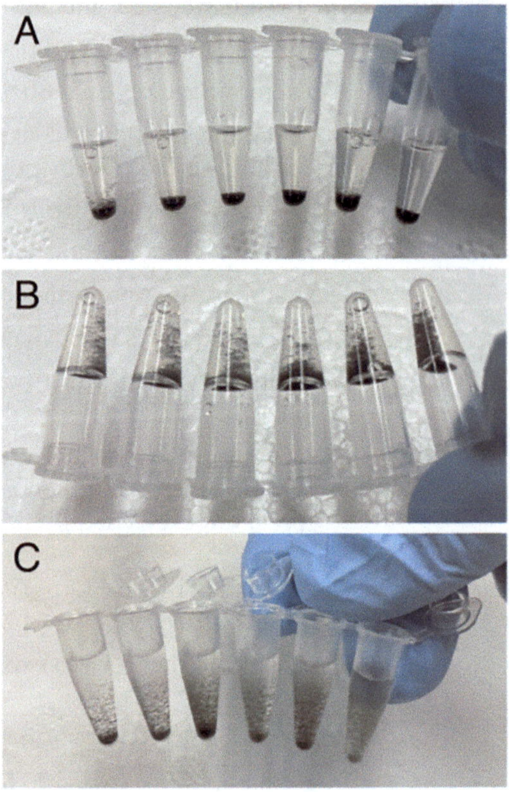

Fig. 2 The behavior of Magne HaloTag Beads. (**a**) and (**b**) Rotation during Halotag purification and DNA affinity purification. Without rotation, the beads settle quickly (**a**). Due to the surface extension, the beads remain in the solution even when the tube or well is inverted (**b**). (**c**) Direct dispensing onto the bead pellet results in an even distribution of beads in the solution. During the settling (right to left), the beads are washed

11. Rotate at 2 rpm at room temperature for 1 h (*see* Fig. 2 and **Note 16**).
12. Briefly spin down and place on DynaMag-96 for 30 s.
13. Discard the supernatant and remove from DynaMag-96.
14. Vigorously pipette 200 μL of PBS-X to ensure the magnetic beads are thoroughly mixed (*see* Fig. 2 and **Note 17**).
15. Place the plate on DynaMag-96 and shift the PCR plate laterally across DynaMag-96, aligning it with each consecutive lane. Since DynaMag-96 has 13 lanes and the PCR plate has 12, this lateral movement allows for the magnetic beads to be exposed to a new magnetic field position with each shift. After moving the plate over one lane, move it back to the original position. Repeat this process ten times to ensure thorough mixing of the solution in each well.

16. Discard the supernatant and remove from DynaMag-96.
17. Repeat **steps 14–16** twice.

3.4 DNA Affinity Purification

1. Dispense 80 μL of PBS-X and 30–100 ng of pre-PCR DNA-seq library from **step 44** in Subheading 3.1 to each well.
2. Mix by pipetting and directly transfer to a new 96-well PCR plate (*see* **Note 18**).
3. Seal the plate with adhesive PCR plate seal (*see* **Note 19**).
4. Rotate at 2 rpm at room temperature for 1 h (*see* Fig. 2 and **Note 20**).
5. Briefly spin down and place on DynaMag-96 for 30 s.
6. Discard the supernatant and remove from DynaMag-96.
7. Repeat **steps 14–16** in Subheading 3.3 four times (*see* **Notes 17 and 21**).
8. Discard the supernatant and remove from DynaMag-96.
9. Dispense 200 μL of PBS-X to each well.
10. Mix by pipetting and directly transfer to a new 96-well PCR plate (*see* **Note 21**).
11. Discard the supernatant and remove from DynaMag-96.
12. Dispense 15 μL of EB to each well.
13. Seal the plate with adhesive PCR plate seal (*see* **Note 19**).
14. Run the DNA recovery reaction as follows (*see* **Note 22**):

    10 min at 98 °C.

    Hold at 4 °C.

3.5 DAP-seq Library Amplification

1. Dispense the library amplification reaction to each well of a new 96-well PCR plate as follows:

KAPA HiFi HotStart ReadyMix	12.5 μL
Unique dual index primer mix (5 μM each)	2 μL
Total	14.5 μL

2. Place the plate from **step 14** in Subheading 3.3 on DynaMag-96 for 30 s.
3. Transfer 10.5 μL of the supernatant to the library amplification reaction from **step 1**.
4. Briefly spin down.
5. Seal the plate with adhesive PCR plate seal (*see* **Note 19**).

6. Run the library amplification reaction as follows:

   2 min at 95 °C

   30 s at 98 °C

   12 cycles of:

   15 s at 98 °C

   30 s at 60 °C

   30 s at 72 °C

   Hold at 25 °C
7. Transfer 5 μL from each well to a single 1.5 mL DNA LoBind tube. Store the remaining library at −20 °C.
8. Add 500 μL of SPRI beads.
9. Rotate at room temperature for 10 min.
10. Briefly spin down and place the tube on DynaMag-2 for 5 min.
11. Discard the supernatant.
12. Add 1 mL of 80% ethanol to the tube and wait 30 s.
13. Discard the supernatant.
14. Repeat **steps 12** and **13**.
15. Briefly spin down and place the tube on DynaMag-2.
16. Remove the residual 80% ethanol as much as possible (*see* **Note 8**).
17. Keep the lid open and air-dry at room temperature for 3 min or until the bead pellet dries (*see* **Note 9**).
18. Remove the tube from DynaMag-2.
19. Add 100 μL of EB and mix well by pipetting.
20. Incubate at room temperature for 5 min.
21. Transfer the supernatant to a new 1.5 mL DNA LoBind tube.
22. Add 100 μL of SPRI beads and mix well by pipetting (*see* **Note 23**).
23. Incubate at room temperature for 10 min.
24. Repeat **steps 10–22**.
25. Measure the concentration with a Qubit Fluorometer and Qubit dsDNA High Sensitivity Assay Kit. Store the remaining DAP-seq library at −20 °C.

3.6 Verification of TF Expression (Optional; See *Notes 4 and 24*)

1. Mix 5 μL of in vitro expression reaction and 5 μL of 2× Laemmli sample buffer.
2. Boil or incubate at 95 °C for 5 min.
3. Assemble the Mini-PROTEAN Tetra Cell and 4–15% Mini-PROTEAN TGX Precast Gel, according to the manufacturer's protocol. Use a roller to remove any bubbles between the filter papers, the gel, and the membrane.

4. Fill the buffer chamber with the electrophoresis buffer, according to the manufacturer's protocol.
5. Load 10 μL of protein samples from **step 2** to each well.
6. Run electrophoresis at 200 V for 30 min.
7. Detach the gel from gel plates and detach gel fingers using a spatula.
8. Perform the electroblotting using a Trans-Blot Turbo Mini 0.2 μm PVDF Packs and a Trans-Blot Turbo, according to the manufacturer's protocol.
9. De-assemble the blotting sandwich and immerse the membrane in TBS-T with gentle agitation for 2 min.
10. Immerse the membrane in 10 mL of blocking solution with gentle agitation for 30 min.
11. Add 1 μL of anti-Halo rabbit primary antibody and gently shake for 1 h.
12. Discard the solution.
13. Add 50 mL of TBS-T and gently agitate the membrane for 10 min.
14. Repeat **steps 12** and **13** twice.
15. Add 10 mL of blocking solution and 1 μL of anti-rabbit IgG, HRP-linked secondary antibody.
16. Gently agitate the membrane for 1 h.
17. Repeat **steps 12** and **13** three times.
18. Develop and detect chemiluminescence using a Clarity Western ECL Substrate and a LAS 4000, according to the manufacturer's protocol.

4 Notes

1. These are required only for optional Subheading 3.5.
2. Required reagents can be purchased separately as NEBNext Ultra II End Repair/dA-Tailing module and NEBNext Ultra II Ligation module.
3. This kit includes NEBNext Adapter for Illumina, USER enzyme, and 96 Unique Dual Index to reduce index hopping. When pooling more than 96 libraries for sequencing, purchase additional oligo sets. Up to 384 unique dual index primers are available.
4. Verification of TF expression by immunoblotting is not mandatory because most TFs can be expressed in this wheat germ system, and it is laborious to perform a lot of immunoblotting. However, if DAP-seq fails, verifying TF expression would help troubleshooting.

5. If the 200 bp target mode is not pre-set, use the following setting: temperature, 20 °C; peak incident power [W], 50; duty factor [%], 20; cycles per burst [cpb], 200; and treatment time [s], 150. Usually, it takes less than 5 min until the water temperature reaches 20 °C.
6. Prepare pre-PCR DNA-seq library from 10 μg of genomic DNA. Recovery rate is approximately 50–70%, resulting in 5–7 μg pre-PCR DNA-seq library, enough for 96 DAP-seq. Since the capacity of microTUBE is 5 μg, split 10 μg of genomic DNA into two microTUBEs.
7. I've observed that the recovery rate is higher when using rotation, compared to the mix-and-place method, especially for larger volumes.
8. The remaining ethanol may lead to a decrease in elution efficiency and inhibit the subsequent enzymatic reactions.
9. Overdrying may lead to a decrease in elution efficiency. If the drying duration exceeds the recommended time, proceed with caution.
10. Sheared DNA fragments have single-stranded and sticky ends. End-prep reaction converts these sticky ends into blunt ends and adds an "A" nucleotide to the 3′ ends. This facilitates efficient adapter ligation.
11. The NEBNext Adapter for Illumina features a unique hairpin loop structure, which includes uracil, to reduce the formation of adapter dimers that can inhibit the PCR reaction. The USER enzyme removes the uracil, transforming the hairpin loop adapter into a Y-shaped adapter.
12. It is hard to thaw TNT SP6 High-Yield Wheat Germ Master Mix on ice or at 4 °C. I thaw it partially at room temperature for 5 min and then place it on ice.
13. During this step, wash and prepare Magne HaloTag Beads in **steps 1–8** in Subheading 3.3.
14. Magne HaloTag Beads are provided as a 20% slurry. When 850 μL is added to the packed beads, the total volume is approximately 1050 μL. Given that Magne HaloTag Beads settle rapidly, ensure to mix them by pipetting prior to each dispensing.
15. It is crucial to prevent PBS-X from touching the upper region of the well. If this happens, the surface tension could be disrupted, causing the beads to get trapped by the adhesive seal during rotation. Recovering beads from the adhesive seal is very difficult.
16. During this rotation, the Halo-tagged TF is captured by the Magne HaloTag Beads. Due to surface tension, the beads can remain in the solution.

17. Direct dispensing onto the bead pellet results in an even distribution of beads in the solution. During the settling, the beads are washed. However, care should be taken to avoid splashing the solution.
18. This transfer step helps to reduce possible non-specific contaminants, which could increase background noise.
19. It is crucial to securely seal to prevent contamination and drying.
20. During this rotation, the Halo-tagged TF captures DNA fragments that contain TF binding sites. Due to surface tension, the beads can remain in the solution.
21. This step is critical to wash away unbound DNA fragments. Do not reduce the number of washes. The final transfer to a new plate helps to reduce potential non-specific contaminants, which could increase background noise.
22. This step denatures the TF and releases the bound DNA fragments from the denatured TF.
23. This second SPRI bead purification ensures the complete removal of primer and adapter dimers.
24. The use of Trans-Blot Turbo system enables to perform the time-consuming blotting in just 3 min. If this step is performed with a subset of TFs, it can be carried out simultaneously with Subheadings 3.3, 3.4, and 3.5 in a day.

Acknowledgments

This work was supported by JSPS KAKENHI grant numbers 23H04756, 23H04968, and 23 K23585.

References

1. Badia-i-Mompel P, Wessels L, Müller-Dott S et al (2023) Gene regulatory network inference in the era of single-cell multi-omics. Nat Rev Genet 24:739–754
2. Schmitz RJ, Grotewold E, Stam M (2022) Cis-regulatory sequences in plants: their importance, discovery, and future challenges. Plant Cell 34:718–741
3. Spitz F, Furlong E (2012) Transcription factors: from enhancer binding to developmental control. Nat Rev Genet 13:613–626
4. Marand AP, Eveland AL, Kaufmann K et al (2023) Cis-regulatory elements in plant development, adaptation, and evolution. Annu Rev Plant Biol 74:111–137
5. Park PJ (2009) ChIP–seq: advantages and challenges of a maturing technology. Nat Rev Genet 10:669–680
6. O'Malley R, Huang SC, Song L et al (2016) Cistrome and epicistrome features shape the regulatory DNA landscape. Cell 165:1280–1292
7. Bartlett A, O'Malley RC, Huang SC et al (2017) Mapping genome-wide transcription-factor binding sites using DAP-seq. Nat Protoc 12:1659–1672
8. Lai X, Stigliani A, Lucas J et al (2020) Genome-wide binding of SEPALLATA3 and AGAMOUS complexes determined by sequential DNA-affinity purification sequencing. Nucleic Acids Res 48:9637–9648

9. Li M, Yao T, Lin W et al (2023) Double DAP-seq uncovered synergistic DNA binding of interacting bZIP transcription factors. Nat Commun 14:2600
10. Ichino L, Boone BA, Strauskulage L et al (2021) MBD5 and MBD6 couple DNA methylation to gene silencing through the J-domain protein SILENZIO. Science 372:1434–1439
11. Galli M, Khakhar A, Lu Z et al (2018) The DNA binding landscape of the maize AUXIN RESPONSE FACTOR family. Nat Commun 9:4526
12. Ricci WA, Lu Z, Ji L et al (2019) Widespread long-range cis-regulatory elements in the maize genome. Nat Plants 5:1237–1249
13. Cerise M, Giaume F, Galli M et al (2020) OsFD4 promotes the rice floral transition via florigen activation complex formation in the shoot apical meristem. New Phytol 229:429–443
14. Kawakatsu T, Teramoto S, Takayasu S et al (2021) The transcriptomic landscapes of rice cultivars with diverse root system architectures grown in upland field conditions. Plant J 106:1177–1190
15. Zhang Y, Li Z, Liu J et al (2022) Transposable elements orchestrate subgenome-convergent and -divergent transcription in common wheat. Nat Commun 13:6940

Chapter 18

High-Efficient Transient Gene Transfection for Transcriptional Activity Assay Using Rice Protoplasts on Large-Scale Platform

Shingo Sakamoto

Abstract

Transient gene expression in protoplasts offers a straightforward and efficient approach for investigating gene function. Here, I introduce a method that utilizes a 96-well plate format to assess transcriptional activation capacity of transcription factors through transient expression system in rice protoplasts. This method includes plant cultivation, protoplast preparation, transfection, and transactivation assay and allows for the simultaneous measurement and validation of 23 samples with 4 replicates per sample at a time.

Key words Protoplast, Transient expression, 96-well plate, Transcription factor, Transactivation assay

1 Introduction

In vivo gene functional analysis is crucial for deciphering gene functions and molecular mechanisms in plants. While stable transformation stands as a prominent method for such analysis, it often involves cell culture processes, which are time-consuming and labor-intensive. In contrast, transient gene expression systems in plant cells or tissues, such as agroinfiltration and direct delivery using a particle gun, offer a more efficient alternative that achieves the desired experimental results in less time and with reduced effort [1–4].

A variety of experimental methods have been developed for transient expression systems using protoplasts, in many plant species, tissues, and evaluation systems [5–8]. These methods are particularly suited for large-scale screening. However, many researchers encounter challenges with instability and poor reproducibility due to the absence of key points in the protocols. Several studies have reported large-scale functional analyses of transcription factors using transcription factor libraries [9], and we have

Taiji Kawakatsu (ed.), *Rice: Methods and Protocols*, Methods in Molecular Biology, vol. 2869,
https://doi.org/10.1007/978-1-0716-4204-7_18,

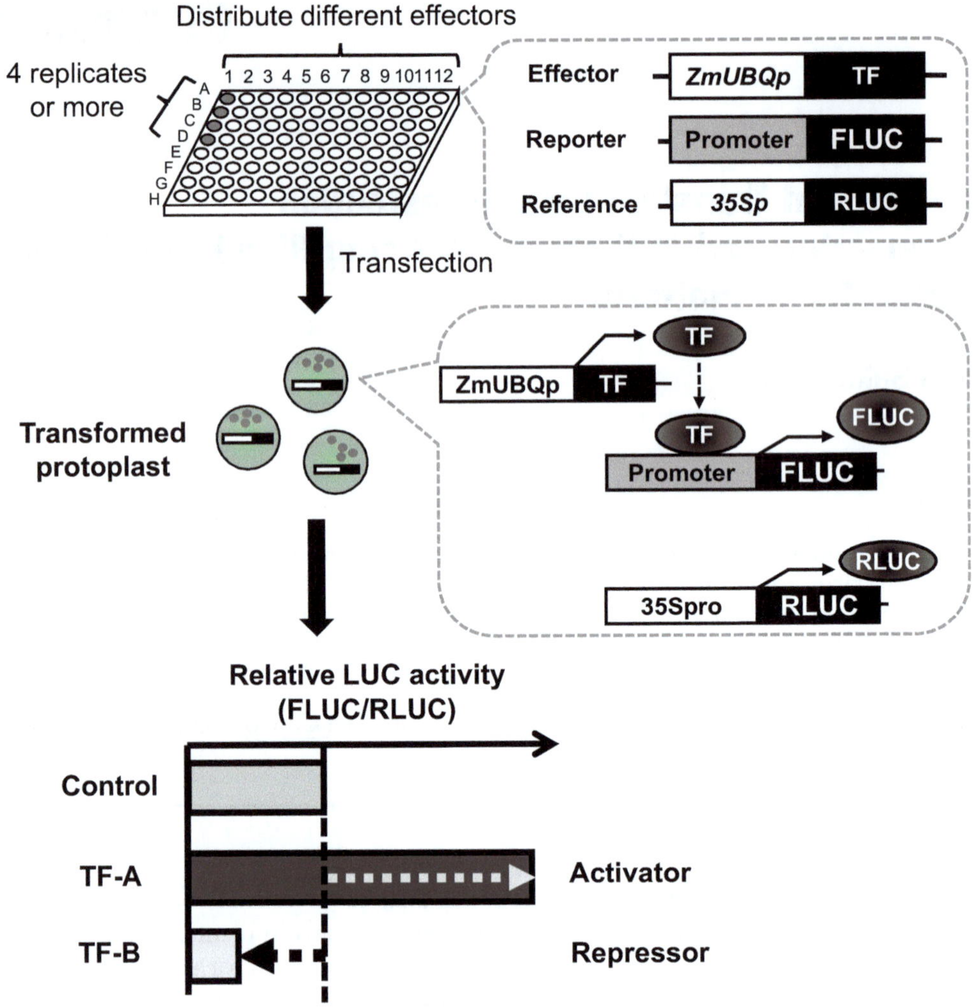

Fig. 1 Schematic diagram of transient transcriptional regulation assay using protoplast cells. Three plasmids (effector, reporter, and reference as internal control) are introduced into protoplasts on 96-well plates. The effectors expressed from the transfected plasmids bind to the promoters on the reporter plasmid, thereby expressing FLUC, and the activities are measured. Apart from that, RLUC is expressed at the constitutive promoter, and its activity is also measured for the purpose of obtaining a correction value, which takes into account the transformation efficiency, cell viability, transcriptional capacity, etc. Both of these measurements are calculated as a relative LUC activity (FLUC/RLUC) and compared with a control to determine whether the transcription factor is an activator or a repressor

successfully established experimental platforms to facilitate such analyses [6, 10] (*see* Fig. 1). This chapter describes plant cultivation, protoplast preparation, transfection, and transactivation assay for transient gene expression method using rice protoplasts [11]. Also, I highlight some of the key elements necessary to obtain stable data for large-scale functional analysis of plant transcription factors.

2 Materials

Prepare all solutions using ultrapure water and analytical grade reagents. Prepare and store all reagents at 4 °C unless indicated otherwise.

2.1 Plant Cultivation

1. Rice seed: *Oryza sativa* L. (cv. Nipponbare) seeds. Store in airtight bottles filled with silica gel at 4 °C (*see* **Note 1**).
2. Culture pot: 232 × 154 × 80 mm (length, width, height) tray with a 1-cm-diameter hole drilled in the bottom. Place the mesh (sink strainer mesh bag, MONOTARO, cat. no. 33656106) at the bottom of the tray to prevent soil leakage (*see* Fig. 2).
3. Culture soil: Bonsol no. 2 (Sumitomo Chemical; *see* **Note 2**).
4. Water tray (*see* **Note 3**).
5. Plastic wrap.

2.2 Protoplast Preparation

1. 200 mM MES, pH 5.7: Weigh 42.65 g of 2-(N-morpholino)-ethanesulfonic acid (MES) and add to a glass beaker containing 900 mL of ultrapure water. Mix and adjust to pH 5.7 with KOH. Make up to 1 L with water. Sterilize and filter using a 0.2 or 0.45 μm filter. Store at 4 °C (*see* **Note 4**).
2. 1 M $CaCl_2$: Weigh 73.51 g of $CaCl_2{\cdot}2H_2O$ and add to a glass beaker containing 400 mL of ultrapure water. Make up to

Fig. 2 Culture pot. Pot with 14 holes of about 1 cm in the bottom. Netting on the bottom to prevent soil from leaking through the holes

500 mL with ultrapure water. Sterilize and filter using a 0.2 or 0.45 μm filter and store at 4 °C.

3. Digestion buffer: 600 mM mannitol, 20 mM KCl, 20 mM MES, pH 5.7, 10 mM $CaCl_2$. Weigh 54.65 g of mannitol and 0.75 g of KCl and add to a glass beaker containing 100 mL of ultrapure water. Add 50 mL of 200 mM MES (pH 5.7) and 5 mL of 1 M $CaCl_2$. Mix well and make up to 1 L with ultrapure water. Sterilize using a 0.2 or 0.45 μm filter. Store at 4 °C.
4. W5 buffer: 150 mM NaCl, 125 mM $CaCl_2$, 5 mM KCl, 2 mM MES, pH 5.7. Weigh 8.77 g of NaCl and 0.37 g of KCl. Add 125 mL of 1 M $CaCl_2$ and 10 mL of 200 mM MES, pH 5.7. Make up to 1 L with ultrapure water. Sterilize using a 0.2 or 0.45 μm filter. Store at 4 °C.
5. MMg solution: 400 mM mannitol, 15 mM $MgCl_2$, 4 mM MES, pH 5.7. Weigh 36.43 g of mannitol and 1.53 g of $MgCl_2 \cdot 6H_2O$. Add 10 mL of 200 mM MES, pH 5.7. Make up to 500 mL with ultrapure water. Sterilize using a 0.2 or 0.45 μm filter. Store at 4 °C.
6. 50 mL conical tube.
7. Cell strainer, 40 μm (Corning 352340).
8. Centrifuge with swing rotor for 50 mL tube.
9. Burker-Turk's counting chamber.

2.3 Transfection

1. 1 M $CaCl_2$: Weigh 73.51 g of $CaCl_2 \cdot 2H_2O$ and add to a glass beaker containing 400 mL of ultrapure water. Make up to 500 mL with ultrapure water. Sterilize and filter using a 0.2 or 0.45 μm filter and store at 4 °C.
2. Polyethylene glycol (PEG)/Ca^{2+} solution: 40% PEG, 200 mM mannitol, 100 mM $CaCl_2$. Weigh 16 g of PEG4000 (Sigma-Aldrich 81,242; *see* **Note 5**) and 1.53 g of $MgCl_2 \cdot 6H_2O$ and add to 50 mL conical tube. Add 4 mL of 1 M $CaCl_2$ and 22.4 mL of ultrapure water (*see* **Note 6**). Mix well by inverting (*see* **Note 7**). Store at room temperature.
3. Plasmid DNA: Extract plasmid DNA using a silica column-based kit. Adjust the concentration to 100–1000 ng/μL (*see* **Note 8**).
4. U-bottom 96-well plate.
5. Centrifuge with swing rotor for 96-well plate.
6. Plate mixer for 96-well plate.
7. Incubator set at 28 °C.
8. Multi-channel electric pipette.
9. Plate seal for 96-well plate.

2.4 Transcriptional Activity Assay

1. Luminometer equipped with dispenser and plate shaking unit (*see* **Note 9**).
2. 96-well plate for luminometer (*see* **Note 10**).
3. Dual-Luciferase Reporter Assay System (Promega; *see* **Note 11**).

3 Methods

3.1 Plant Cultivation

1. Immerse about 200 rice seeds in 2 L water and incubate at 28 °C in the dark (*see* Fig. 3a and **Note 12**).
2. Replace with fresh water every 12 h and repeat for 2 days.
3. Fill a culture pot with 1 L of culture soil and place it on a water tray.
4. Add water to the water tray to absorb water into soil just before use.
5. Choose germinated seeds from **step 1** and arrange approximately 70 seeds (7 vertical × 10 horizontal) on the soil (*see* Fig. 3b, c).
6. Place the culture pot with the germinated seeds on the water tray in the incubator set at 28 °C.
7. Add enough water to the culture pot to fully submerge the seeds, and then cover the culture pot with plastic wrap.
8. Incubate at 28 °C in the dark to grow for 2 days.
9. Remove the plastic wrap, add water to the water tray, and continue to cultivate for an additional 3–5 days.

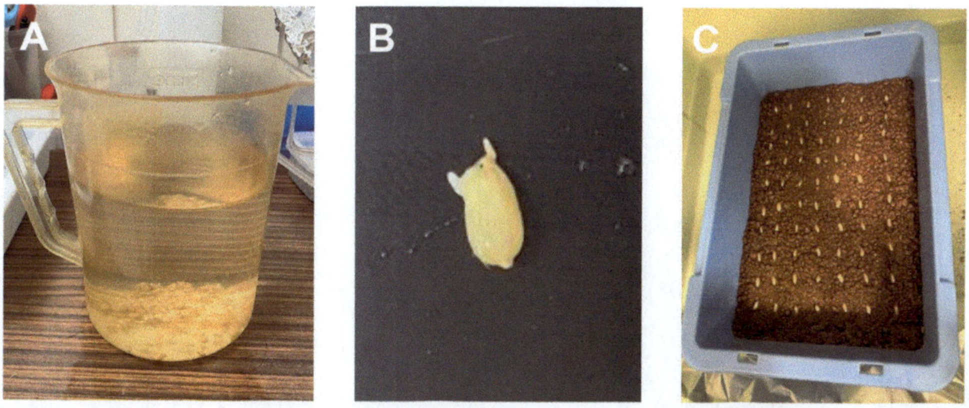

Fig. 3 Rice seed germination and cultivation. (**a**) Seeds absorbing water. (**b**) Germinating rice seeds 2 days after water absorption. A little shoot at the top. Seeds in this condition are selected from seeds that have absorbed water. (**c**) Germinated seeds are placed in the culture medium in such a way that they are slightly covered. The seedlings are then filled with enough water to bring all the culture medium below the water surface, covered with a light-shielding lid and grown at 28 °C for 5–6 days

3.2 Protoplast Preparation

1. Prepare the enzyme solution in a 50 mL conical tube. Suspend 0.20 g of Cellulase Onozuka RS, 0.10 g of Macerozyme R-10, and 0.02 g of BSA in 20 mL of digestion buffer.
2. Add 14 μL of 2-mercaptoethanol to the enzyme solution and mix well.
3. Apply a vacuum for 30 s to eliminate air bubbles present in the enzyme solution suspension. Perform this step a total of five times.
4. Cut culms of 20 rice plants at 5–6 days after sowing vertically with a new scalpel (*see* Fig. 4a, b). Instead of pushing down to cut, slide the blade to slice them into 1 mm pieces and soak them in the enzyme solution (*see* Fig. 4c).
5. Apply a vacuum for 1 min and release to the normal pressure. Perform this step a total of three times to facilitate the penetration of the enzyme solution into the tissue.
6. Incubate at 25 °C with shaking at 80 rpm for 2 h.
7. Place a 40 μm cell strainer on top of a new 50 mL conical tube.
8. Pour the solution in the reaction tube from **step 6** through the strainer into the collection tube from **step** 7.
9. Strain the remaining protoplast in the residual tissues by pressing them against a filter using a mounting side of 200 μL pipette tip.
10. Collect the residual protoplasts in the reaction tube. Add 20 mL of W5 buffer to the reaction tube and gently mix by inverting (*see* **Note 13**). Pour the solution through the strainer into the collection tube.
11. Centrifuge with swing rotor at 200× *g* for 10 min.

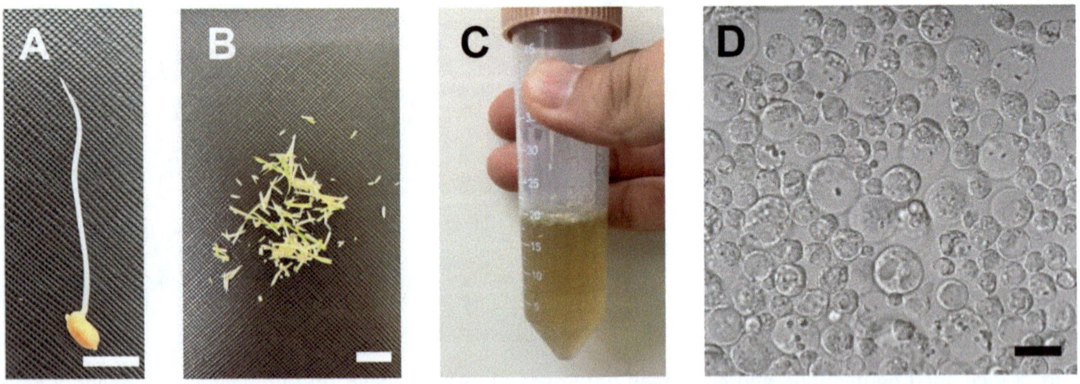

Fig. 4 Protoplast isolation from etiolated rice seedlings. (**a**) Young rice sheath on days 5–6, being sliced diagonally to about 1 mm. (**b**) Sliced rice seedlings. (**c**) Sliced shoot in an enzyme solution. (**d**) Isolated protoplasts. The cells are generally 5–10 μm in diameter. Scale bar indicates 1 cm for (**a**) and (**b**) and 20 μm for (**d**)

12. Remove the supernatant.
13. Add 20 mL of W5 buffer to the pellet and suspend gently. Do not form bubbles.
14. Repeat **steps 11–13** three times.
15. Incubate on ice for 10 min (*see* **Note 14**).
16. Suspend the settled protoplasts gently and centrifuge with swing rotor at 200× *g* for 5 min.
17. Remove supernatant and resuspend in 20 mL of MMg solution.
18. Centrifuge at 200× *g* for 5 min.
19. Remove the supernatant and suspend the pellet in 5 mL of MMg solution.
20. Measure the concentration of protoplasts using a Burker-Turk's counting chamber, according to the manufacturer's protocol.
21. Dilute the protoplast suspension to 2.0–2.5 × 10^5 cells/mL with MMg solution.

3.3 Gene Transfection

1. Prepare 10 μL of LDNA solutions containing 100 ng of the effector plasmid DNA, 500 ng of the reporter plasmid DNA, and 10 ng of reference plasmid DNA.
2. Dispense 10 μL of the DNA solution into 96-well plate.
3. Dispense 40 μL of the protoplast solution (8–10 × 10^6 cells) into 96-well plate.
4. Dispense 50 μL of PEG/Ca^{2+} solution into 96-well plate.
5. Mix by vortexing. Begin at a speed of 900 rpm, escalate to 1700 rpm, hold for 5 s, revert back to 900 rpm, and then stop (*see* **Note 15**).
6. Incubate at room temperature for 10 min.
7. Vigorously add 200 μL of W5 buffer to mix by a water flow.
8. Centrifuge with swing rotor at 200× *g* for 5 min.
9. Remove 200 μL of the supernatant.
10. Add 200 μL of W5 buffer vigorously.
11. Centrifuge with swing rotor at 200× *g* for 3 min.
12. Remove 200 μL of the supernatant.
13. Repeat **steps 10–12** three times.
14. Seal the plate with a plate seal and incubate at 28 °C in the dark overnight (*see* **Note 16**).

3.4 Transcriptional Regulation Assay

1. Prepare LAR II and Stop & Glo reagent according to the manufacturer's protocol.
2. Set LAR II and Stop & Glo reagent to injectors 1 and 2 of the luminometer.
3. Dispense 20 μL of 5× passive lysis buffer into each well.
4. Shake the suspension at 900 rpm on a plate shaker for 15 s.
5. Incubate at room temperature for 10 min to allow the cells to lyse.
6. Dispense 5 μL of the cell lysate from **step 3** into the 96-well white plate for luminometer.
7. Set the plate dispensed with the cell lysate into the luminometer.
8. Assess the FLUC activity by dispensing 20 μL of LAR II from injector 1. Mix the solution using the instrument's stirring function for 5 s, and then measure the luminescence in each well for 1 s. This measurement is recorded as Score A, representing the activity score.
9. Assess the RLUC activity by dispensing 20 μL of Stop & Glo reagent from injector 2. Mix the solution using the instrument's stirring function for 5 s, and then measure the luminescence in each well for 1 s. This measurement is recorded as Score B, representing the internal control score.
10. The activation score is calculated by dividing Score A by Score B. Compare this value with those obtained when a control effector is introduced (*see* Fig. 1 and **Note 17**). If the activation score in the experimental group exceeds that in the control group, the effector transcription factor is deemed a transcriptional activator. Conversely, if it is lower, the effector transcription factor is considered a transcriptional repressor.

4 Notes

1. Rice cultivars other than Nipponbare work as well.
2. Any culture soil for rice seedlings can be used.
3. We use a water tray that accommodates two culture pots (**item 2**).
4. Do not autoclave MES buffer. Although the exact reason is unknown, the use of autoclaved MES has been observed to cause this method to fail. All subsequent solutions containing MES buffer should not be autoclaved and instead sterilized through filtration.
5. The use of this specific PEG is essential. Otherwise, the experiment may fail.

6. Upon completion of the mixing process, the resulting solution will have a total volume of 40 mL.
7. PEG and mannitol are difficult to dissolve. Mix the solution by inversion at least 10 min.
8. The reporter is constituted with a gene of interest promoter driving FLUC. The effector is a transcription factor driven by the constitutive *Zea mays UBIQUITIN* (*ZmUBQ*) promoter, and the internal control is RLUC driven by the constitutive cauliflower mosaic virus *35S* promoter. A terminator should be connected 3′ downstream of each gene. The terminator of *Arabidopsis HEAT SHOCK PROTEIN 18.2* (HSPter) is preferred.
9. Any measurement method is acceptable if it can measure both the measured value and the correction value as an internal control.
10. Transparent plates should be avoided as the emitted light can interfere with readings in adjacent wells.
11. This protocol uses the dual reporter assay system from Promega, but systems from other manufacturers can also be used. Additionally, alternative methods, such as a combination of firefly luciferase and GUS activity, which is a fluorometric assay, can be employed.
12. Although 20 rice plants can yield enough protoplasts for a single 96-well plate assay, it is better to start with a larger quantity of seeds. This allows for the selection of seeds that germinate synchronously and can be transplanted into the soil.
13. Keep W5 buffer chilled on ice.
14. This step can be omitted without affecting the results.
15. If the solution is not mixed adequately, increase the shaking speed to approximately 1700 rpm.
16. In transcriptional regulation assays, the results are typically stable between 12 and 18 h, depending on the experiment.
17. Depending on the experimental design, it is recommended to obtain at least four or more replicates.

Acknowledgments

The author thanks Dr. Yuji Teramura (AIST), Dr. Tomoko Niki (AIST), Ms. Fumiko Hosaka (AIST), and Ms. Manami Masuda (AIST) for critical advice and their skillful technical assistance. This work was partially supported by the JSPS KAKENHI grant numbers 19 K16174 and 23 K23678 to S.S.

References

1. Abel S, Theologis A (1994) Transient transformation of Arabidopsis leaf protoplasts: a versatile experimental system to study gene expression. Plant J 5:421–427
2. Gilliard G, Huby E, Cordelier S et al (2021) Protoplast: a valuable toolbox to investigate plant stress perception and response. Front Plant Sci 12:749581
3. Godon C, Caboche M, Daniel-Vedele F (1993) Transient plant gene expression: a simple and reproducible method based on flowing particle gun. Biochimie 75:591–595
4. Kapila J, De Rycke R, Van Montagu M et al (1997) An agrobacterium-mediated transient gene expression system for intact leaves. Plant Sci 122:101–108
5. Sakamoto S, Fujikawa Y, Esaka M (2013) Analysis of ascorbic acid biosynthesis using a simple transient gene expression system in tomato fruit protoplasts. Biosci Biotechnol Biochem 77:673–675
6. Sakamoto S, Matsui K, Oshima Y et al (2020) Efficient transient gene expression system using buckwheat hypocotyl protoplasts for large-scale experiments. Breed Sci 70:128–134
7. Yang Y, Shao Y, Chaffin T et al (2023) Synthetic promoter screening using poplar mesophyll protoplast transformation. Bio-Protocol 13:e4660
8. Wroblewski T, Tomczak A, Michelmore R (2005) Optimization of agrobacterium-mediated transient assays of gene expression in lettuce, tomato and Arabidopsis. Plant Biotechnol J 3:259–273
9. Wehner N, Hartmann L, Ehlert A et al (2011) High-throughput protoplast transactivation (PTA) system for the analysis of Arabidopsis transcription factor function. Plant J 68:560–569
10. Sakamoto S, Nomura T, Kato Y et al (2022) High-transcriptional activation ability of bamboo SECONDARY WALL NAC transcription factors is derived from C-terminal domain. Plant Biotechnol 39:229–240
11. Kitomi Y, Hanzawa E, Kuya N et al (2020) Root angle modifications by the DRO1 homolog improve rice yields in saline paddy fields. Proc Natl Acad Sci USA 117:21242–21250

Index

A

Acid phenol/chloroform/isoamyl alcohol 16–19
Acrylamide 30, 32, 33, 37, 71
Adaptor ligation 136
Affinity tag purification 148
Agarose 16, 18, 19, 51, 77, 94, 96, 97, 99, 105, 107, 108, 114, 117, 137, 143, 144, 149, 153, 154, 158, 161
Agencourt Chloropure 80, 84, 94, 98, 99, 105, 109
Agrobacterium-mediated transformation 75–78, 88, 94, 98, 105, 109
Agrobacterium tumefaciens 78
Alkaline phosphatase (AP) 50, 62, 64, 65, 70, 71
Ammonium persulfate (APS) 30, 32, 33
Ampicillin 63, 78, 79, 88, 105, 107, 154
Antibody
- primary antibody
  - anti-digoxigenin-AP 52, 55
  - anti-Halo rabbit antibody 172
  - anti-OSH1 rabbit antibody 159, 162
  - anti-OSH1 rabbit antiserum 62, 64
- secondary antibody
  - anti-rabbit IgG, HRP-linked 172, 179

Antisense probes 50, 57
Aqueous phase 19, 139
Argonaute (AGO) 135, 136
Aspirator 19, 22, 23, 25, 64, 66, 72

B

Base editors
- adenine base editors (ABEs) 101–103, 105, 107, 110
- cytosine base editors (CBEs) 101–103, 105, 107, 110

Base substitutions 101–110
Bead mill homogenizer 9, 10, 12, 16, 17, 136
Binary vectors 77, 82, 94, 96, 105, 108
Bioinformatic analysis
- BEDtools 138, 142
- bowtie 115, 120
- BWA 141
- fastp 115, 120
- fastQC 137, 140
- RSEM 115, 120
- samtools 115, 120, 138, 142
- Shortstack 138, 141
- Trim galore 137, 140
- Trimmomatic 137, 140
- UMI-tools 115, 120

Bovine serum albumin (BSA) 30–32, 53, 55, 64, 65, 149, 152, 159, 162, 165, 188
Bradford assay 37

C

Chromatin extraction 157, 158, 160
Chromatin immunoprecipitation followed by sequencing (ChIP-seq) 166, 169, 170
Circadian clocks 29–38
Cleavage Under Targets and Release Using Nuclease (CUT&RUN) 147–155
Colony PCR 86, 87
CRISPR-Cas systems
- CRISPR/Cas9 91–100
- CRISPR-Cas12f 76

D

Deaminases
- cytidine deaminase 102, 103
- deoxyadenosine deaminase (TadA) 102

Desiccator, *see* Vacuum desiccator
Diethylpyrocarbonate (DEPC) 16, 18, 19
Digoxigenin (DIG) 50
DNA affinity purification followed by sequencing (DAP-seq) 170, 172, 173, 176–180
DNA double-strand breaks (DSBs) 92
Drought stress 1
Dual-Luciferase reporter assay system 187

E

Electroblotting 30, 33, 34, 38, 63, 179
Electrophoresis 16, 18, 19, 30, 33, 57, 58, 63, 71, 77, 79, 80, 84, 86, 88, 94, 96, 97, 99, 105, 107, 108, 110, 139, 149, 150, 152, 153, 158, 161, 172, 179
Electroporation 78, 82

Taiji Kawakatsu (ed.), *Rice: Methods and Protocols*, Methods in Molecular Biology, vol. 2869,
https://doi.org/10.1007/978-1-0716-4204-7,

Embryos 21–26, 88, 89
Epigenomes 147, 170
External RNA Controls Consortium
(ERCC) 125, 127

F

Fluorescein Isothiocyanate (FITC) 62, 64, 65, 68, 70, 72
Formaldehyde crosslinking 157

G

Gene expression 1, 2, 7, 15, 41, 113, 135, 147, 169, 183, 184
Gene regulatory network (GRN) 169
Gene targeting (GT) 91–100
Genome editing 75, 76, 91–93, 101, 109
Genomic DNA 10–12, 25, 66, 84, 98, 102, 109, 141, 149, 153, 154, 166, 171–173, 180
Graded ethanol series 22, 23, 41, 43, 44, 52, 64, 68, 71
Guanidine thiocyanate (GTC) 7–9, 12

H

Halo-tag 170, 171, 175, 176
Heteroduplex mobility assay (HMA) 84, 88
Histochemical staining, *see* Histochemistry
Histochemistry 41–48
Histology 41–48
Histone modification 50
Homologous recombination (HR) 91
Hybridization stringency, *see* Hybridization temperature
Hybridization temperature 58
Hydroponic culture 1–4
Hygromycin-resistant calli 84, 98, 99, 109

I

Immunoblotting 29–38, 165, 166, 179
Immunodetection 34
Immunohistochemical staining (IHC) 61, 62, 64, 70
Immunological detection, *see* Immunodetection
Immunoprecipitation 157, 158, 162, 169
In situ hybridization 41, 42, 46, 49–51
In vitro expression 170, 175, 178

L

Laser capture microdissection (LCM) 21–26
Lithium chloride precipitation 15, 54

M

Magnetic stand 9, 10, 12, 116, 117, 128, 129
Micrococcal nuclease (MNase) 149, 152, 155
Mutation analysis 80, 84–87, 105, 109, 110
Mutation detection 84

N

Next generation sequencing (NGS) 123, 124, 136, 138, 141, 143
Northern blot 16

O

Osmotic stresses 1–4

P

PAG-MNase 147–155
Paraffin embedding 22, 23, 41, 64, 66
Paraffin sectioning 62
Positive-negative selection 91, 92
Protein-DNA interactions 170
Protein electroblotting system 34, 63
Protein electrophoresis system 63
Protein extraction 37, 63, 66
Protein quantification 29, 31
Protein separation, *see* SDS-PAGE
Protoplasts 183–185, 188, 189, 191

Q

Quantitative PCR (qPCR) 7, 12, 118, 119, 130, 141, 157, 163, 167
Quantitative RT-PCR 16

R

Recombinant protein purification 148
Restriction enzymes 54, 57, 77, 94, 105, 107, 108, 110
Reverse transcription (RT) 7, 12, 114–117, 121, 123–128, 131, 133, 135–144
Ribosomal RNA (rRNA) 13, 26, 124
RNA degradation 12, 19, 21, 22, 25, 120, 124
RNA extraction 7–13, 15–19, 23, 24, 113
RNA integrity 12, 13, 19
RNA integrity number (RIN) 11, 13, 26, 145
RNA probes 49–51, 54, 55, 58
RNA-seq
- DeLTa-seq 113–121
- Lasy-Seq 123–133
- Targeted RNA-seq 113–121
- 3' RNA-seq 113, 124

S

SDS-PAGE 63, 67, 71, 149–152, 154, 155, 165
Sectioning, *see* Paraffin sectioning
Single guide RNA (sgRNA) 76, 77, 87–89, 92–94, 96, 97, 99, 102–105, 107, 108, 110
Small interfering RNAs... 13, 140
Small RNA extraction 8, 17
Solid-phase reversible immobilization (SPRI) beads 8
Spectinomycin 78–80, 82, 88, 105, 108
Starchy endosperms 15, 17
Stem-loop RT-PCR 135–143

T

Targeted mutagenesis 75–77, 83, 84, 87, 89
Total RNA extraction 17, 22
Transactivation assay 184
Transcription factor (TF)
  binding sites 169–181
  DNA interaction 170
  heterodimer-DNA interactions 170
  methylated DNA interactions 170
Transcriptional activity assay.. 187
Transfection 184, 186, 189
Transient expression.. 183

V

Vacuum desiccator 43, 55
Vacuum pump 43, 158
Vector construction..77, 78, 80, 93–98, 102, 104–108, 110
Vector transformation 65

X

X-ray films.. 36
Xylene 42

Y

Y-shape adapter 126, 129, 130, 132, 180

MIX
Papier aus verantwortungsvollen Quellen
Paper from responsible sources
FSC® C105338

If you have any concerns about our products,
you can contact us on
ProductSafety@springernature.com

In case Publisher is established outside the EU,
the EU authorized representative is:
Springer Nature Customer Service Center GmbH
Europaplatz 3, 69115 Heidelberg, Germany

Printed by Libri Plureos GmbH
in Hamburg, Germany